Human–Animal Relationships in Times of Pandemic and Climate Crises

This book situates sociological research as a vital tool for understanding, and responding to, the multispecies entanglements that cause, inform and arise from states of crisis involving the environment, climate and zoonotic disease transmission. Considering the consequences of a range of multispecies engagements that challenge the perceived distinction between the social worlds of humans and other animals, it explores the themes of crisis through a range of studies, including ecological disturbance, consumer culture, intensive farming and interspecies relations in urban life. With attention to central questions about life in 'the now normal', including the extent to which a human–animal perspective can contribute to our understanding of pandemics, the ideological foundations of mainstream norms for human–animal relations and the scope of current and emerging social movements for reshaping human–animal relations, this volume represents a timely and important call for a sociological vision to embrace the implications of a multispecies planet and to expand the concepts of inclusion and justice. A reconsideration of the human–animal relation that seeks both to revise sociology's past and inform its future, *Human–Animal Relationships in Times of Pandemic and Climate Crises* will appeal to scholars across the social sciences with interests in human–animal relations and the environment.

Josephine Browne is a sociologist who has most recently held positions at Griffith and Southern Cross Universities, where she brings critical approaches to animal studies and gender in her teaching and research.

Zoei Sutton is a lecturer in Sociology at Flinders University pursuing critical, non-human animal-centric research.

Multispecies Encounters
Series editors:

Samantha Hurn is Associate Professor in Anthropology, Director of the Exeter Anthrozoology as Symbiotic Ethics (EASE) working group and Programme Director for the MA and PhD programmes in Anthrozoology at the University of Exeter, UK.

Chris Wilbert is Senior Lecturer in Tourism and Geography at the Lord Ashcroft International Business School at Anglia Ruskin University, UK.

Multispecies Encounters provides an interdisciplinary forum for the discussion, development and dissemination of research focused on encounters between members of different species. Re-evaluating our human relationships with other-than-human beings through an interrogation of the 'myth of human exceptionalism' which has structured (and limited) social thought for so long, the series presents work including multi-species ethnography, animal geographies and more-than-human approaches to research, in order not only better to understand the human condition, but also to situate us holistically, as human animals, within the global ecosystems we share with countless other living beings.

As such, the series expresses a commitment to the importance of giving balanced consideration to the experiences of all social actors involved in any given social interaction, with work advancing our theoretical knowledge and understanding of multi-species encounters and, where possible, exploring analytical frameworks which include ways or kinds of 'being' other than the human.

Published

Animal Lives and Why They Matter
Arne Johan Vetlesen

Cat People
Human-Cat Interrelatedness in the Cat Fancy
Emily Stone

The Giant Squid in Transatlantic Culture
The Monsterization of Molluscs
Otto Latva

Human–Animal Relationships in Times of Pandemic and Climate Crises
Multispecies Sociology for the New Normal
Edited by Josephine Browne and Zoei Sutton

Human–Animal Relationships in Times of Pandemic and Climate Crises

Multispecies Sociology for the New Normal

Edited by Josephine Browne and Zoei Sutton

Routledge
Taylor & Francis Group

LONDON AND NEW YORK

First published 2025
by Routledge
4 Park Square, Milton Park, Abingdon, Oxon OX14 4RN

and by Routledge
605 Third Avenue, New York, NY 10158

Routledge is an imprint of the Taylor & Francis Group, an informa business

© 2025 selection and editorial matter, Josephine Browne and Zoei Sutton; individual chapters, the contributors

British Library Cataloguing-in-Publication Data
A catalogue record for this book is available from the British Library

ISBN: 978-1-032-18039-7 (hbk)
ISBN: 978-1-032-19148-5 (pbk)
ISBN: 978-1-003-25791-2 (ebk)

DOI: 10.4324/9781003257912

Typeset in Times New Roman
by SPi Technologies India Pvt Ltd (Straive)

Contents

Illustrations

Figures

Tables

Contributors

Josephine Browne teaches social and political science at Southern Cross University. She was previously research fellow at Griffith University, Australia. Her research focuses on human–animal relationships and masculinities, particularly in contemporary culture and literature. A former social welfare worker and a narrative therapist, she first researched human–animal relations in the mid-1990s, founding Australia's earliest national service for bereaved humans of animal companions (*Agape*) as a result. She serves on the Australasian Animal Studies Association (AASA), as co-convener of The Australian Sociological Association (TASA) Sociology and Activism thematic group and on The Australian Women's and Gender Studies Association (AWGSA) and is associate editor for the journal *Feminisms, Gender and Advocacy*. Josephine's current research is with Dr Chantelle Bayes, on more-than-human subjectivities. She can be found on Bluesky @drejosephine. bsky.social.

Matthew Cole is a lecturer in Criminology at The Open University in the United Kingdom. He has been writing about veganism since the mid-2000s and has particular research interests in the discursive representation of non-human animals, the (mis)representation of veganism and the development of vegan sociology, vegan criminology and multidisciplinary vegan studies. He co-wrote his first book with Dr Kate Stewart, investigating the socialisation of speciesist and anti-vegan norms in childhood: *Our Children and Other Animals: The Cultural Construction of Human and Animal Relations in Childhood* (2014).

Erika Cudworth is a senior lecturer in the School of Applied Social Sciences at De Montfort University, Leicester, United Kingdom. Her research interests include complexity theory, gender and human relations with non-human animals, particularly theoretical and political challenges to exclusive humanism and the development of post-humanist ethnography. She is the author of *Environment and Society* (2003), *Developing Ecofeminist Theory* (2005) and *Social Lives with Other Animals* (2011); co-author of *The Modern State* (2007), *Posthuman International Relations* (2011) and *The Emancipatory Project of Posthumanism* (2018); and co-editor of *Technology, Society and*

Inequality (2013) and *Anarchism and Animal Liberation* (2015). Erika's current projects are a book on people's relations with dog companions and a critical reappraisal of the legacy of classical social theory for post-humanist scholarship.

Heather Fraser is an associate professor and coordinator of the Master of Social Work Program at Queensland University of Technology. She is a critical social worker whose work focuses on love but also violence and abuse of humans and non-human animals. Visual and narrative methods, and animal-assisted and art interventions, are her other specialities.

Lynda M. Korimboccus has been a committed ethical vegan and grassroots activist since 1999. She is a passionate advocate for equity and justice, and her doctoral research investigates speciesism and the lived experiences of young vegan children in key Scottish social institutions such as education. Lynda holds an MA in anthrozoology as well as undergraduate honours degrees in philosophy, politics, social psychology and sociology. She is an active member of The Vegan Society's Researcher and Education Networks and a volunteer writer for *Faunalytics*. Writing independently in the field of critical animal studies and vegan sociology, Lynda is also Editor-in-Chief of the International Association of Vegan Sociologists' *Student Journal of Vegan Sociology* and has taught sociology at West Lothian College, Scotland for more than 15 years. She can be found on Twitter @lmkorimboccus.

Catherine Oliver is a lecturer in the Sociology of Climate Change at Lancaster University, United Kingdom. Catherine's research interests include veganism, chickens, urban studies and more-than-human theory. She previously worked as a postdoctoral researcher at the University of Cambridge, where she was researching city chickens in London, ex-commercial hens and the galline Anthropocene. Catherine completed her PhD in 2020 at the University of Birmingham on veganism in Britain. Her first book, *Veganism, Archives, and Animals*, was published by Routledge in 2021. Catherine has published widely on veganism, vegan geographies, chickens and other animals, and her work can be found on Twitter @katiecmoliver and on her website, https://catherinecmoliver.com.

Zoei Sutton is a lecturer in Sociology at Flinders University pursuing critical, non-human animal-centric research. She is the co-founder of the International Association for Vegan Sociologists (IAVS), co-created and co-convenes The Australian Sociological Association's (TASA) Sociology and Animals thematic group and is a member of the Institute for Critical Animal Studies (ICAS) Oceania collective. She currently serves as Book Review Editor for *Society & Animals* and is an affiliate of the New Zealand Centre for Human Animal Studies (NZCHAS). Her dissertation examined the lived experience of human–companion animal entanglements, utilising species-inclusive methods to centre non-human animals' experiences and encourage critical reflection on them. Other recent projects have examined the construction of killable 'pest' species in print media and the depoliti-cised treatment of other non-human animals in the sociological literature and print media. Underlying this research is a deep commitment to research that is *for* non-human animals, rather than merely *about* them. She can be found on Twitter @zoei_sutton.

Nick Pendergrast has a PhD in Sociology and his thesis applied sociological theories on social movements and organisations to the animal advocacy movement in both Australia and the United States. He teaches Sociology at the University of Melbourne and has published widely on critical animal studies, human–non-human relations, the media, social movements and social change.

Nik Taylor is a critical and public sociologist whose research focuses on mech-anisms of power and marginalisation expressed in/through human relations with other species; it is informed by critical/intersectional feminism. Nik is currently Co-Director of the New Zealand Centre for Human Animal Studies and a professor at the University of Canterbury, where she teaches topics that focus on human–animal violence links, scholar advocacy, social change and feminism, crime and deviance – particularly domestic violence and animal abuse. Nik's latest books include *Queer Entanglements* (with Damien Riggs, Shoshanna Rosenberg and Heather Fraser, 2021) and *Companion Animals and Domestic Violence: Rescuing Me, Rescuing You* (with Heather Fraser, 2019).

Dinesh Wadiwel is an associate professor of Socio-Legal Studies and Human Rights in the Discipline of Sociology and Criminology, University of Sydney. Dinesh's research interests include theories of violence, critical ani-mal studies and disability rights. He is the author of *The War Against Animals* (2015), co-editor (with Matthew Chrulew) of *Foucault and Animals* (2016) and his essays have appeared in *Cultural Studies Review*, *Angelaki*, *New Literary History* and the *South Atlantic Quarterly*. Dinesh's new mon-ograph, *Animals and Capital*, was released in 2023.

Corey Lee Wrenn is a lecturer of Sociology in the School of Social Policy, Sociology and Social Research (SSPSSR) and Co-Director of the Centre for the Study of Social and Political Movements at the University of Kent, United Kingdom. She received her PhD in sociology from Colorado State University in 2016. She was awarded Exemplary Diversity Scholar, 2016 by the University of Michigan's National Center for Institutional Diversity. She served as a council member of the American Sociological Association's Animals and Society section (2013–16), was elected chair in 2018, and co-founded the International Association of Vegan Sociologists in 2020. She serves as a Book Review Editor for *Society & Animals* and Editor for *The Sociological Quarterly*, is a member of the Vegan Society's Research Advisory Committee and hosts the *Sociology & Animals Podcast*. Dr Wrenn has published in several peer-reviewed academic journals including the

Journal of Gender Studies, Environmental Values, Feminist Media Studies, Disability & Society, Food, Culture & Society and *Society & Animals*. In July 2013, she founded the Vegan Feminist Network, an academic-activist project engaging intersectional social justice praxis. She is the author of *A Rational Approach to Animal Rights: Extensions in Abolitionist Theory* (2016), *Piecemeal Protest: Animal Rights in the Age of Nonprofits* (2019) and *Animals in Irish Society* (2021).

Acknowledgements

We would like to acknowledge the custodians of the lands we live, think and write on – for Josephine, the Yugambeh people of Bundjalung Country, for Zoei, the Kaurna people. We recognise these are unceded and sovereign lands that always were and always will be, Aboriginal lands. We pay respects to Elders past and present, acknowledging histories of violence that remain to be addressed on these lands.

This book wouldn't exist without the encouragement and support of Neil Jordan, Alice Salt and Gemma Rogers at Routledge. Thank you for seeing the value in this work and for your patience as we put this project together amidst the disruptive context of pandemic and other crises.

We are grateful for the continued support of The Australian Sociological Association for our work: in particular, we acknowledge that this volume emerges from our research panel for TASA 2020. Warm thanks to those on this panel, and to all the additional contributors to this volume, who have shared their research work. Special thanks to Sally Daly at TASA, who has been endlessly helpful to our Sociology and Animals Thematic Group, the Association more broadly for funding part of this research and to colleagues with whom we have shared many rich conversations and ideas along the way.

We also appreciate the support and contributions of colleagues from the International Association for Vegan Sociologists (IAVS). From our inaugural conference, online and international, in 2020, the association continues to grow and be a rich, vibrant, challenging and stimulating forum for our ideas, research formulation and writing. May this association grow and thrive in the years ahead – the insights and challenges are sorely needed in our damaged world as we hope to demonstrate in the following pages.

We acknowledge financial support from the Griffith Centre for Social and Cultural Research, Griffith University, where Josephine benefitted from a decade of colleagial and other support, a time during which this research was made possible. Particular thanks are also due to our flexible and understanding editor, Susan Jarvis.

Josephine wants to thank all her contributing colleagues, whose commitments to bettering human-animal relations continue to inspire her. Thank you also to Zoei Sutton, whose enthusiasm led me to the Sociology and Animals and IAVS communities. I first met Zoei at the Australasian Animal Studies Association (AASA) conference in Christchurch in 2018, and she has since inspired my work on the usefulness of sociological theory for furthering multispecies justice. I am proud to have worked with her on this volume. I wish also to acknowledge my precious animal-oriented children, who deeply care about critical animal research – may your future contain ever-increasing justice for all species and beings. This research path would never have arisen without the immeasurable grace of the companionship of rabbits and cats who have lived with me, teaching me much that continues to challenge and strengthen my research; special thanks to Ben, the rabbit who began it all.

Zoei would like to acknowledge the generosity and patience of the non-human animals who share their lives with her. While your presence comforts, teaches and enriches my life, I know that you in turn must tolerate my (too many) hours on the dreaded computer, encroachment on your garden spaces and general bumbling as I try to create pockets of space that facilitate your 'beastly places' but am all too aware that I often fall short in the face of ever-encroaching Anthroparchy. To my vibrant and generous colleagues working across non-human animal inclusive fields – your ideas, companionship and dissatisfaction with the anti-animal status quo inspire me always. Thanks to the authors in this book who committed their time, support and work to this project and whose constant dedication to taking critical animal scholarship forward is much appreciated. Thank you, Josephine, for your commitment to pushing this book forward, patience with my delayed edits and constant belief in the worth of this project and the need for a book like this, especially in our current times. Finally, thank you to my partner who is a constant source of support and has joined me in this (perhaps sometimes strange seeming) commitment to de-centring humans in our home and family as much as possible.

Introduction

Introducing multispecies sociology for the new normal

Zoei Sutton and Josephine Browne

Chapter

Should the new normal be shaped to differ from the old?
Henry A. Wise Wood (1918, p. 604)

In the early months of the coronavirus lockdown in South Australia, a footage of a solitary grey kangaroo hopping through the deserted central business district (CBD) was shared in a tweet by the local police: 'Protective Security Officers tracked a suspect wearing a grey fur coat hopping through the heart of the Adelaide CBD this morning. He was last seen on foot heading into the West Parklands' (South Australian Police, 19 April 2020).

Bounding down once-familiar streets, then unrecognizable in the absence of noise, traffic and people, the presence of this furry 'city slicker' served to remind us not only of our own disappearance from these cityscapes but also of the life that went on when there were no humans around to deter it. For many, these moments were bittersweet – observed by humans from spaces of confinement where they were already acutely aware of their own absence from these public sites. Examples of free-living animals visibly occupying public spaces previously claimed by humans came to be known as the 'animal takeover', as mentioned in the above tweet. Some claims – for instance, of dolphins swimming through the Venice canals – were found to be fictitious. However, there were documented sightings of deer, boars, raccoons, wild turkeys, pumas and other animals venturing further into 'human' spaces as the movements of the human world were increasingly restricted over the COVID-19 pandemic (Singh, 2020). For some non-human animals, these ventures into spaces considered to be 'human' areas were new – enabled by our species' retreat from shared public life (Vardi, Berger-Tal and Roll, 2021). For others, attention to the 'animal takeover' was perhaps more indicative of the fact that humans were suddenly paying much more attention to their environment and those who inhabited it (Vardi, Berger-Tal and Roll, 2021). Regardless, the COVID-19 global pandemic served to visibilize the often-overlooked multispecies nature of the social world, from zoonoses to consumption of flesh and companionship and interactions

DOI: 10.4324/9781003257912-1

with 'nature'. This awareness is sorely needed, and yet, early hopes that the pandemic might lead to a true reckoning with the implications for structural and epistemological changes for other animals consequent on the crisis were not realised (Arcari, 2021). An increased understanding of the implications of multispecies relationships is required not just in the context of the pandemic but also of the other disasters we currently face, including climate change and the accompanying intensification of natural disasters.

When we decided to create this book, Australia was in the recovery period of the Black Summer bushfires, a catastrophic event estimated to have killed approximately three billion animals and burned 18.6 million hectares of land (van Eeden et al., 2020) and approaching the end of the second year of the COVID-19 pandemic. We wanted to write a book that not only brought together current experts in multispecies sociologies to visibilize what human–animal entanglements are like in the social world now but also to uncover what lessons might be learned as we consider possible future world-making in the wake of these and other disasters. The persistence of calls to 'return to normal', particularly in the early days of the pandemic, is contrasted with an increasingly urgent need for a greater response to the climate crisis (IPCC, 2023), underlining that 'normal' was and is problematic, 'that the underlying causes of the present ecological turmoil are inseparable from the entire configuration of social, cultural, political and economic patterns' (Savransky, 2022, p. 371). Multispecies sociology is deeply invested in critiques of both pandemics and climate crises, in re-examining the ongoing and increasing human–animal interfaces that provide the opportunity for cross-species disease and in the many threats, deaths and extinctions to more-than-human worlds concurrent with human suffering in climate crises. The remainder of this chapter will explore the two main themes – the new normal and multispecies sociology – in greater depth before concluding with an overview of the structure of the volume.

The new normal: Pandemics, disasters and climate change

Although its origins have been attributed to everything from terrorist attacks to the global financial crisis in 2008 and even the Y2K bug, the opening quote of this chapter shows that humans have been discussing 'the new normal' since at least the end of World War I (if not long before). It has been defined as the 'dramatic economic, cultural and social transformations that caused precariousness and social unrest, impacting collective perceptions and individual lifestyles' (Corpuz, 2021, p. e344). The term has now been used to describe the dramatic shifts in social life brought about by COVID-19, such as lockdowns, isolation, mask-wearing, hand-sanitizer, disruption to services and working from home (for those employed in jobs where this was and is a possibility). Less frequently mentioned are the over-burdened health systems, loss of life and emotional toll of living in constant awareness of death, disease and uncertainty. The disruption to social life has been immense, but it is by no means the only force disrupting the social order at present.

As we mentioned earlier, when we began this book, Australia was post-bushfire and in the second year of COVID-19. At the time of writing, one of us was dealing with 'unprecedented' floods brought about by a storm cell intensified by the heating of the Pacific Ocean – a mere five years after the previous 'one in 100 years' flooding event in the region. Twenty-two humans had died, alongside thousands of non-human animals. The COVID-19 pandemic had continued into a third year, with vaccine inequity seen to be a major challenge for global health and recovery. Monkeypox had emerged as a new zoonotic disease around the globe. Climate change continues to intensify changes in weather and accompanying natural disasters, leading to more frequent and intense floods, fires, hurricanes and other catastrophic events (IPCC, 2022). From our perspective, the new normal is not a post-event transformation, but rather a reckoning with the dramatic and continuing crises that are inextricably intertwined with how humans relate to other animals and the natural environment. In response to Wise Wood's quote at the beginning of this Introduction, we would say that the new normal must be dramatically different from the world as it is now, and deliberately fashioned to involve living 'less badly' with other animals. In this, we turn to Cudworth and Hobden's (2018, p. 148) approach to social transformation, in which they argue that 'Our emancipation comes not through positive transformation towards a clear future "good", but in moving away from living badly – we need to be living "a less wrong life"'. We contend that multispecies sociology is well placed to inform this move away from 'living badly'.

The promise of multispecies sociology

Sociology as a discipline has not always been inclined to include non-human animals within the bounds of the social world it purports to study. The early history of the discipline included concerted efforts for its recognition as a scientific, objective study of the human social world (Bernardeau Moreau, 2015), at a time when inclusion of animal-others, particularly from a critical perspective, remained associated with emotionalism, women and the non-rational. Influenced by a binaristic Cartesian understanding of a human–animal divide that positioned humans as intellectually and emotionally superior, early sociological studies additionally dismissed other animals as lacking the agency to meaningfully interact with the social world (Mead, 1934; Peggs, 2012). Calls to attend to this anthropocentrism and seriously consider non-human animals were largely ignored until the late 1970s (Bryant, 1979; Taylor and Sutton, 2018); however, the development of sociological animal studies as a field did not truly begin until around the 1990s.

As it stands today, multispecies sociology – in which we include all sociological works that meaningfully include non-human animals in their focus – is well placed to inform the reimagination of a 'less bad' new normal. The merit of these works lies in their capacity to 'show' the complex and problematic ways in which humans are currently entangled with non-human others, and in doing so, highlight the anthropocentric 'bad' that must be challenged if we are to reckon with the current crises (Cudworth and Hobden, 2018; Thaler, 2021). Multispecies sociology draws on the sociological canon, which excels in its capacity to visibilize power relations and the connections between broader social structures and everyday life, and adapts these theories to extend their usefulness to the multispecies world in which we all live. In doing so, sociologists have usefully contributed to understandings of the intersecting systems of domination that shape human and non-human lives (e.g. Cudworth, 2011; Nibert, 2013; Chapter 7 of this volume), the agency and symbolic interaction of non-human animals (e.g. Alger and Alger, 2003; Carter and Charles, 2013) and the social construction of non-human animals according to their dis/utility to humans (e.g. Cole and Stewart, 2014; Sutton and Taylor, 2019), to name just a few areas. As will be seen in this volume, the field also offers tremendous promise in providing new ways to understand current multispecies relations on a global scale (Chapter 8), consider overlooked theoretical approaches (Chapter 3), learn from and with aligned activist movements (Chapters 4, 5 and 6) and engage with futures thinking utilizing utopian and dystopian imaginaries to remake the world (Chapter 10). There is immense promise in this burgeoning field, and an urgency for the possibilities it provides, given pandemic and climate catastrophes increasing 'the escalating evidence that life on the planet is facing a system-wide catastrophe' (Cudworth and Hobden, 2018, p. 1). The chapters in this volume provide an introduction to the avenues through which we, as a society, might begin to challenge, unthink and rebuild our shared social world in the new normal so that human–animal relations are no longer impeding our reckoning with current crises, and neither are we ignoring that other animals also live in and are affected by these crises induced by humans (Thaler, 2021).

Why multispecies sociology, and why now? After the shock of the first weeks of the Covid pandemic, emerging voices saw parts of this upset as an opportunity. Growing frustration at inadequate or outright ignoring of scientists' insistence that key 'tipping points' in planetary stability were rapidly approaching, requiring urgent action, had received inadequate response from governments. School strikes for climate, which had grown so rapidly, bolstering a belief in change for the sake of future generations, were receiving less attention. At the same time, climate catastrophes were increasing in frequency and intensity, as scientists had warned (IPCC, 2022). Covid, it seemed, in necessitating social isolation for millions, shutting down cities and airports, was the global wake-up call we might finally heed. Discussion of the word 'normal' – a nostalgic throwback or a necessary change? – became common. In large part, the 'promise' in which C. Wright Mills might have connected individual lives with global pandemic catastrophe remains under-realized (Mills, 1959). Without such a connection, Mills argued, we are doomed to perpetuate the normal, to fail in sociology's great aim of critical examination with the implication of effective social change (Mills, 1959). When considering sociology's history of attention to both power and social change, it is imperative that sociologists engage with this opportunity for global change. Asymmetrical power relations perpetuated by the dualisms relevant to the many possible micro, meso and macro enmeshments of humans and other animals have a great deal to offer critical geopolitical, natural and environmental issues arising from pandemics and climate catastrophes. For instance, the animal–industrial complex as a nascent site of species' cross-contamination, and a source of disease, was highlighted during Covid. Yet, the critique failed to extend to a critical examination of the history of this normalized industry, where the vulnerability of all animal species becomes shared. At the same time, the killing of other species at an industrial scale in order to supply a constructed commodification remained unexamined, as has the 'meat' industry's large contribution to gas emissions, even within the sociology of climate crises (Twine, 2023).

An investment in social change is present throughout the discipline's history, including many sociologists who have provided support and critique of their own communities' efforts, for example, Du Bois and Cox's work as both sociologists and Civil Rights Activists (Lewis, 2014) and Collins and Davis's foundational work in Black feminist activism and sociology (Allen, 2021). Analysis and information proceed, and often attend and bolster, attention to expanding notions of justice: unacknowledged or denied interconnections in complex social systems remain unchallenged, and it is the sociologist's task 'as a producer of knowledge to assess and respond to adverse conditions' (Allen, 2021, p. 34). Leveraging opportunities for change in response to these conditions, a key issue is in de-naturalising a range of practices, languages and socio-cultural conditions under which now-normative arrangements have emerged, thus challenging dominating assumptions of social issues 'as ahistorical, or not open to change' (Twine, 2023, p. 110).

In considering the justice claims for/of other species, the task of sociological advocates of expanding equality and addressing power becomes more pressing and more complex, limited as it is by a persistent human-centric focus on language. Useful analysis has been done on forms of other-animal resistance, which underlines resistance to current norms (e.g. Wadiwel, 2016), but the language argument relies on not just this denial but a range of denials constructed through dualistic logic, including the human as animal, the necessary conditions for life and the ramifications of global climate shifts on all species. While these are macro considerations, the micro and meso analyses of sociology enable richer responses to proposals for change, such as localized food production, Indigenous-led land management practices and significant geopolitical change recognising the place of sustainability and care as enabling greater health and well-being, including restoration and recovery, for futures that are indeed 'less bad'. This collection, then, demonstrates the range and scope of possible directions a multispecies sociology contributes. This includes rethinking languages of conceptualization, structures of dominating power (including among sociologists), social norms of human–animal interactions within and around homes and original theoretical approaches that challenge us to reimagine the complex and multifaceted links between the 'new normal' and key sociological ideas. This collection points to the perspectives and possibilities of considering other animals as a group or class of communicating Others, frequently powerless and agentic-denied beings who are subjected to marginalized, commodified or disregarded lives within dominating human-centric Western paradigms.

The dominating Western paradigm of extraction and a constructed separation from what is termed 'nature' has failed to recognize the reality of a finite biosphere. Enormous global and localized effort is required to meet the demands of both shifting social norms to futures that recognize 'the importance of intergenerational equity and environmental sustainability…[at] the top of the political agenda and to the core of personal and societal belief systems' (McAlpine et al., 2015, p. 56), and to multifaceted structural solutions. The Sixth Assessment Report of the IPCC indicates scientific consensus that the 1.5-degree limit agreed in Paris in 2015 is now unlikely (IPCC, 2023). Despite some efforts toward this goal, government willingness, globally, has stymied adequate action and extensive climate impacts are now being experienced in every region: scientists warn that the next decade's decisions will determine planetary health for thousands of years (IPCC, 2023). Given the extent of increasing threats, the contributions of multispecies sociologists are more relevant and necessary than ever. The connections and disconnects between human exceptionalism, disease spread, environmental destruction and human–animal relationships are all deeply implicated in the research of multispecies sociologists, and their work has numerous possibilities for restructuring and building new ways forward into futures of greater justice and care. These analyses, collectively, provide us with the capacity to learn to live 'less badly' alongside other species today, as well as with the structural critiques necessary to inform broader changes in the global sociopolitical and environmental landscapes.

A guide to the book

The book is structured into three parts. Part I: Animals in Everyday Life brings together three chapters that explore sociological research with non-human animals in the everyday life of the 'new normal'. Each chapter emphasizes the entanglement of real, living animals in relations of domination with the human species – whether as 'pets' or lively commodities existing for human pleasure, 'farmed' animals imprisoned in food systems or the 'objects' of study in research that is not adequately designed to engage with multispecies lives. The authors emphasize the need to look closely and engage critically with the lived materialities of non-human animals' lives, remaining mindful that 'our' practices and methods can easily reinscribe anthroparchal power relations even if our intentions are good.

In Chapter 1, Catherine Oliver explores the interspecies relations between chickens and those who kept them during the COVID-19 pandemic. Drawing on interviews with urban chicken-keepers, she argues that the surge in chicken-keeping during the pandemic did not lead to more equitable relations between species, but instead reproduced and heightened the oppressive expectations of labour on domestic hens. However, in paying close attention to these everyday interspecies entanglements, we can also glean an understanding of how some humans are choosing to live differently with other animals, transforming relationships and perceptions of animals and the city itself.

This attention to the reproduction of inequitable relations is continued in Chapter 2 by Zoei Sutton, who explores media representations of companion animals in Australia during COVID-19. She demonstrates that news stories overwhelmingly emphasize the commodity status of animals used as pets and focus on human-centric perspectives of these relationships. Criticizing the commodification of minded beings for lively companion 'goods', Sutton calls for a re-examination of multispecies companionship in which humans might consider how to be better companions to the non-human animals with which we already cohabit in our homes.

Chapter 3, by Nik Taylor and Heather Fraser, challenges the often human-centric methods used to produce knowledge of multispecies worlds. They argue that humans as creators of knowledge products are in a position of power and that this power has been used to maintain the exclusion or marginalization of other animals in the sociological field. As an alternative, they suggest a productive synergy between sociology and posthumanism in order to produce 'less bad' explorations of animals in everyday life.

Part II: Activism explores current sociological research on animal advocacy. The chapters included in this section consider how veganism and explicitly pro-animal stances are received into the social world, and the impact of this on pro-animal humans and advocacy efforts. Each author focuses on a different aspect of the social world for animal advocates which together provide insights into trends in resistance to veganism and animal advocacy and pathways forward to better support pro-animal stances.

In Chapter 4, Lynda M. Korimboccus explores the everyday socialization of speciesism for children, arguing that the anti-animal status quo hides within the barriers of normativity. Drawing on sociological thought around social class, gender and interactionism, Korimboccus highlights the contradictory narratives encountered by children as they are encouraged to feel affection for animals through activities that often rely on animal exploitation and calls for the need to nurture compassion and cruelty-free ethics in children in order to secure a better future for all.

In Chapter 5, Nick Pendergrast presents an analysis of the framing of two animal advocacy campaigns in Australia to explore how current advocacy groups are approaching this work. His analysis identifies examples whereby advocacy efforts focus on individual change rather than speaking to structural drivers of animal exploitation. Contemplating the causes of animal exploitation, Pendergrast argues for the need for a structural approach in advocacy efforts, highlighting productive ways forward for pursuing social change.

In Chapter 6, Corey Wrenn reviews British media representation of veganism to identify trends in attitudes towards pro-animal stances and vegans generally. Reflecting on prior studies in which the British media were found to be largely 'vegaphobic', Wrenn argues that change is evident and the mediascape might now be trending towards a supportive 'vegan curious' stance.

The concluding Part III: Species(ist) Relations includes four chapters that make key theoretical contributions to reimagining human–animal relations by highlighting speciesist and complex multispecies entanglements in the current 'new normal'.

In Chapter 7, Dinesh Wadiwel critically employs a Marxist lens to articulate the structural position of non-human animals under capitalism. Through a deep engagement with conflict theories and Marxist approaches as they have been applied to animals, Wadiwel reinvigorates this key element of the sociological canon to show that we must critically engage with 'food' animals as an economic class. The implications of this exploration highlight a pathway forward for advocates seeking to challenge this domination of non-human animals and the hierarchical anthropocentrism that accompanies it.

In Chapter 8, Erika Cudworth introduces bordering theory as a way of understanding dominant responses to COVID-19 and the exacerbation of intersected forms of inequality brought by the pandemic. She challenges us to instead consider a post-humanist approach to reimagining the new normal by challenging the current ordering of species that sees Western humans as superior and able to engage in the killing, displacement and exploitation of non-human others. Without such a radical shift, the new normal continues to look very familiar indeed, and pandemics are likely to keep emerging.

Also speaking on the familiarity of the 'new normal', in Chapter 9 Matthew Cole explores the construction of non-human animals as 'killable' for food post-Brexit. Cole draws on the socio-discursive construction of animals for human use, demonstrating how species considered to be unpalatable may be renamed and reconstructed in order to make them suddenly desirable to the flesh-eating consumer. In tracking these discursive construction practices across a variety of sources, Cole demonstrates that the media portrayals demonstrate the embeddedness of speciesism in our society, and in doing so highlight both the sites and difficulties of challenging these norms.

In the final chapter of the collection, Chapter 10, Josephine Browne presents a critical analysis of the fictional work *The Animals in That Country*. Drawing on the sociological understandings of dystopia, Browne highlights how literature can be used to both reflect and inform the social world, arguing that texts such as this can move anthropocentric paradigms. She concludes by arguing that dystopic works are performing important functions in the social world, including directing attention to the place of multispecies biography in the sociological imagination. The sociocultural response to these works indicates that sociological fiction, and dystopic works specifically, may increase public concerns for the environment and human treatment of other animals.

Taken together, these works demonstrate the power of multispecies sociology to visibilize our current human–animal entanglements and expose sites where humans could indeed be considered to be 'living badly' with non-human others. The lessons learned from this examination of our own species can inform a less-oppressive, less-bad and less-human-centric 'new normal' that moves towards multispecies flourishing and a social world more prepared to address the pandemic and environmental crises of our times.

References

Alger, J. and Alger, S. (2003). *Cat Culture: The Social World of a Cat Shelter*. Philadelphia, PA: Temple University Press.

Allen, S.E. (2021). The Black-Feminist roots of scholar-activism: Lessons from Ida B. Wells-Barnett. In Z. Luna and W. Pirtle (eds), *Black Feminist Sociology: Perspectives and praxis*. Routledge: New York.

Arcari, P. (2021). The Covid pandemic, 'pivotal' moments, and persistent anthropocentrism: Interrogating the (il)legitimacy of critical animal perspectives. *Animal Studies Journal*, 10(1), 186–239.

Bernardeau Moreau, D. (2015). Intervention sociology: History and foundations. *International Review of Sociology*, 25(1), 180–93.

Bryant, C. (1979). The zoological connection: Animal-related human behavior. *Social Forces*, 58(2), 399–421.

Carter, B. and Charles, N. (2013). Animals, agency and resistance. *Journal for the Theory of Social Behaviour*, 43(3), 322–40.

Cole, M. and Stewart, K. (2014). *Our Children and Other Animals: The Cultural Construction of Human–Animal Relations in Childhood*. Aldershot: Ashgate.

Corpuz, J.C.G. (2021). Adapting to the culture of 'new normal': An emerging response to COVID-19. *Journal of Public Health*, 43(2), e344–45.

Cudworth, E. (2011). *Social Lives with Other Animals: Tales of Sex, Death and Love*. Basingstoke: Palgrave Macmillan.

Cudworth, E. and Hobden, S. (2018). *The Emancipatory Project of Posthumanism*. London: Routledge.

IPCC. (2022). *Climate Change 2022: Impacts, Adaptation, and Vulnerability – Contribution of Working Group II to the Sixth Assessment Report of the Intergovernmental Panel on Climate Change*. Cambridge: Cambridge University Press.

IPCC. (2023). *AR6 Synthesis Report: Climate Change 2023*. Cambridge: Cambridge University Press.

Lewis, L.F. (2014). The iconoclasm of Oliver Cromwell Cox and the critique of white supremacy. *Canadian Journal of Latin American and Caribbean Studies*, 39(3), 345–63.

McAlpine, C.A., Seabrook, L.M., Ryan, J.G., Feeney, B.J., Ripple, W.J., Ehrlich, A.H. and Ehrlich, P.R. (2015). Transformational change: Creating a safe operating space for humanity. *Ecology and Society*, 20(1), 56.

Mead, G.H. (1934 [1962]). *Mind, Self and Society*. Chicago: University of Chicago Press.

Mills, C.W. (1959). *The Sociological Imagination*. Oxford: Oxford University Press.

Nibert, D. (2013). *Animal Oppression and Human Violence: Domesecration, Capitalism, and Global Conflict*. New York: Columbia University Press.

Peggs, K. (2012). *Animals and Sociology*. Basingstoke: Palgrave Macmillan.

Savransky, M. (2022). Ecological uncivilization: Precarious world-making after progress. *The Sociological Review Monographs*, 70(2), 367–84.

Singh, M. (2020, March 22). Emboldened wild animals venture into locked-down cities worldwide. *The Guardian*. www.theguardian.com/world/2020/mar/22/animals-cities-coronavirus-lockdowns-deer-raccoons 4/10/2022

South Australian Police. (2020, April 19). Protective security officers tracked a suspect wearing a grey fur coat hopping through the heart of the Adelaide CBD this morning. *Twitter*. https://twitter.com/sapolicenews/status/1251721467189813250?lang=en

Sutton, Z. and Taylor, N. (2019). Managing the borders: Static/dynamic nature and the 'management' of 'problem' species. *Parallax*, 25(4), 379–94.

Taylor, N. and Sutton, Z. (2018). For an emancipatory animal sociology. *Journal of Sociology*, 54(4), 467–87.

Thaler, M. (2021). What if: Multispecies justice as the expression of utopian desire. *Environmental Politics*, 31(2), 258–76.

Twine, R. (2023). Where are the nonhuman animals in the sociology of climate change? *Society & Animals*, 31(1), 105–130.

van Eeden, L., Nimmo, D., Mahony, M., Herman, K., Ehmke, G., Driessen, J., O'Connor, J., Bino, G., Taylor, M. and Dickman, C. (2020). *Australia's 2019–2020 Bushfires: The Wildlife Toll Interim Report*. Sydney: WWF Australia.

Vardi, R., Berger-Tal, O. and Roll, U. (2021). iNaturalist insights illuminate COVID-19 effects on large mammals in urban centers. *Biological Conservation*, 254, 108953.

Wadiwel, D.J. (2016). Do fish resist? *Cultural Studies Review*, 22(1), 196–242.

Wise Wood, H.A. (1918). Beware! *National Electric Light Association Bulletin*, 5(11), 604.

Part I
Animals in everyday life

1 The chicken city

Urban interspecies sociabilities

Catherine Oliver

Introduction

During the COVID-19 pandemic and subsequent lockdown restrictions in Britain, backyard chicken keeping saw a huge surge, with live chickens selling out, according to the BBC (Griffith, 2020). This was because (middle-class) people now had the time to look after chickens due to working from home, and a perceived food scarcity at the sight of empty supermarket shelves resulted in people wanting to feel in control of their food sources.

Keeping chickens was not as easy as many novice keepers believed, however. People who had taken in chickens were reported as abandoning them or leaving foxes to 'take care of them' (Mellen, 2020). The surge in chicken keeping didn't lead to more equitable relations between chickens and humans but instead was reproducing and even heightening the oppressive expectations of labour on domestic hens (Oliver, 2021a). At the same time, the situation for chickens in industrial agriculture worsened significantly during the pandemic. As reported by the National Farmers' Union on 12 June 2021, shortages of supplies for chickens, such as bedding, food, and fuel, as well as labour shortages, meant poultry farms' operations were slowed. Rather than this proving beneficial for chickens, it left them open to being without even the bare minimum of care for months while all farm inspections were halted.

The pandemic has changed things for some chickens, a trend that has also been reflected in the United States (Chappell, 2020) and Australia (Gaffney and Kinninment, 2020). Yet, while this trend for keeping backyard hens seems to have experienced a pandemic boom, domestic keeping has been on the rise for at least a decade, with urban keeping representing a substantial subsection of this (Blecha and Leitner, 2014). Urban chicken keepers do not keep chickens 'simply to save money or to pursue an eccentric hobby, but rather as an explicit effort to promote and enact alternative urban imaginaries' (Blecha and Leitner, 2014, p. 86). These alternative urban imaginaries create space for new interspecies sociabilities in the city. The social practices of urban chicken keeping in London are the focus of this chapter, which argues that the backyard chicken challenges the boundaries of the urban social milieux.

DOI: 10.4324/9781003257912-3

This chapter is structured into four sections. First, I detail my research methods. In the second section, I draw on interviews to show how and why people began keeping chickens in London. In the third section, I discuss issues of class, gender, chickens, and the city. London is one of the most unequal cities in the world: while the super-rich buy up property as speculative investments (Atkinson, 2020), the city has become hostile to the urban poor, making living with animals in the city a classed and gendered social relationship. In the final empirical section, I focus on three urban interspecies sociabilities: (1) human–chicken sociabilities; (2) multispecies sociabilities beyond the chicken; and (3) the expansion of local community. Finally, I draw conclusions about what the boom of the chicken city means for a 'new normal' and its role in rethinking the urban multispecies milieux.

I draw on interviews with urban chicken keepers with a range of perspectives, from liberation to utility relations, not to show how urban flocks are spaces of liberation or absent from violence but rather to argue that the increase in urban chickens presents an opportunity to understand how humans are choosing to live differently with animals and how this transforms the perceptions of 'farmed' animals and of the city itself.

Methods

In July 2021, I began work on ethnographic fieldwork with London's chickens, which took place across various community chicken keeping spaces. Despite this research beginning a year earlier, I was neither able to visit my fieldwork city nor to spend time with any chickens for 12 months. In lieu of ethnographic work, 20 interviews with urban chicken keepers living in London were conducted via video call. These were semi-structured interviews lasting between one and two hours, which were transcribed and coded manually. The research in this chapter was overshadowed by the restrictions that affected all researchers across the world during the pandemic (Oliver, 2022). Nonetheless, through online interviews, this research has captured how the pandemic allowed more people than ever before to open their lives and homes to chickens.

London's chicken boom

In this section, I explain the two major ways by which chickens are brought into the city – the rehoming of commercial laying hens and the breeding of specialist chickens – before detailing why people decide to keep chickens in the city.

Getting chickens into the city

One major way of getting chickens into the city is to buy them through specialist chicken sellers, who themselves buy imported or bred birds from chicken breeders across the world. Two chicken sellers, Jean and Peter,[1] who run a chicken supplier business near Heathrow Airport explained to me how they started their business:

> We had just bought five chickens and lost them to a fox. We struggled to buy locally, so we had to go to Oxfordshire or Sussex. My daughter was really into chickens, so I said, 'why don't we sell some chickens?' We bought 10 and sold them within a week. Then we bought 30 and sold them within a month. We're in a good location here. That's how it started. Before we knew it, we were getting chickens in every few weeks and selling out. For the past ten years – our supplier just retired a couple of years ago – he was supplying us with 100 chickens every four weeks. We were selling about 1200 chickens a year.

Located on the periphery of London, Jean and Peter's business took off alongside the growing interest in urban chicken keeping, saving London chicken keepers from driving to Oxfordshire or Sussex. Jean and Peter's business was recommended to me by multiple people as the place where they bought chickens in London. Chicken selling and breeding, especially on a small scale, has limited regulations in Britain. For example, while large commercial breeders have welfare and vaccination standards for their chickens (APHA, 2019), Jean and Peter told me that vaccination is not a requirement for hobby breeders and that vaccines are usually not seen as necessary in small flocks as disease risk is low (*The Chicken Vet*, 2021). For people keeping fewer than 50 birds, there is no need to register the flock (DEFRA, 2018), making it difficult to keep track of how many domestic chickens are living across Britain.

A second way in which people find birds is through chicken rehoming. Karabozhilova et al. (2012) report that between 2005 and 2012, some 200,000 hens were rehomed across Britain. Five per cent of these were in London. It is much easier to trace the flows of chickens through rehoming organizations, which keep extensive records of their operations. The huge surge of interest in backyard chicken keeping can be traced rudimentarily through these numbers; in 2019 alone, one organization rehomed 60,000 hens.

The rehoming process begins at chicken farms. Rehoming organizations work with farmers to purchase their flocks at about 18 months old. Generally, only female chickens (hens) are rehomed, although there are a small number of exceptions. Rehoming organizations work closely with poultry farms, building long-standing relationships to collect thousands of birds to rehome. Teams of volunteers transfer birds from farms to rehoming distribution centres. The chickens are then collected by keepers on the same day, reducing the stress for the birds. Several of my interviewees had rehomed chickens through these organizations, with three of them going on to work as volunteers. Rich told me that he and his wife:

> started keeping chickens about five years ago. When we moved to a house, one of the things that I wanted was to have two or three chickens. I was originally going to get some purebred chickens. We knew about rescuing

ex-factory chickens, but I didn't really want to do that, because I'd heard that they're hard to look after and die quickly, and that you'll spend every day putting Vaseline on their backs. Then we saw some videos of chickens being outside for the first time. We just thought it was lovely, it's amazing to rescue them. So we got ex-battery chickens from this rehoming place. Over the years, we've began volunteering with that place.

Other urban keepers had previously bought chickens from chicken sellers, but then turned to rehoming birds. For example, mother and daughter Sarah and Fiona told me that in 2013, when Sarah was 13, they first got chickens from a breeder before finding out about hen rehoming:

Fiona: We started with two purebreds. We had Betty,[2] who was a bluebell – and she only passed away in the summer just gone. We had Spaghetti, they were the first two, and Spaghetti lived for four years. She was a Maran Cuivree. Both were lovely, pretty hens. They were from a breeder …

Sarah: Somewhere near Cobham, but I can't remember where it was exactly. Was it Horsley?

Fiona: We started off with two, then one died, and we got two more, so that was four.

Sarah: Two rescue hens that was!

Fiona: Yes, since we had the first hens, I learnt about rescue hens, as I didn't know about them at the time. Since then, all we have had are rescue hens, and we've had 11 hens in total.

A third group of people kept a mix of rehomed and specialist hens, and a fourth small group of interviewees kept only specialist breeds of hens, with one of my interviewees breeding hens for chicken shows. Purebred birds are more typically bought from sellers in the edges of London, or by travelling out to the home counties, where there is more space. Similarly, rehoming organizations tend to operate from further afield, with sites in Essex and Milton Keynes, for example, meaning that birds are almost always brought into the city in (trans) national flows rather than being bred there.

Alternative imaginaries: Keeping chickens as an urban intervention

For many people, chicken keeping is closely entangled with 'caring' visions. Varying kinds of care were on display across the 20 interviews, ranging from those viewing birds as lively commodities (Collard and Dempsey, 2013) to those with total liberation vegan ethics who viewed their urban flocks as 'micro-sanctuaries' (Microsanctuary Resource Centre, 2021). Across these differing perspectives, there were shared urban imaginaries at play, akin to Blecha and Leitner's (2014, p. 14) findings that urban keepers wanted to turn 'the backyard to urban barnyard'. In my interviews, some elements of this vision were present, but even more pressing was a desire to bridge the rural and urban to bring a bit of 'the good life' into the city (Oliver, 2021a).

Urban chicken keeping is often related to concerns over where our food comes from and its impact on the world. While only one person confided that they killed and ate their backyard chickens,[3] almost all understood chickens as 'pets with benefits' (Gaffikin, 2009). However, the hens' eggs were more than just food; for the keepers, the backyard egg symbolizes a transformed relationship with chickens and, through this, with food systems more widely. The surge in backyard chicken keeping is closely related to the rise of movements such as locavorism (Barnhill, 2015) and ideas of moving towards (usually unrealized) self-sufficiency. The production of the egg realizes these human values, through both human and non-human labour (Barua, 2017) in the urban barnyard.

In London, the entanglement of chickens with environmental and ethical concerns takes on a further layer of salience. At least since the late eighteenth century, (farmed/food) animal life has emptied out of London as Britain's food production was expanded in the colonial hinterlands and the city itself was sanitized (Otter, 2020). In the intervening centuries, London has been oriented towards the humans (and increasingly towards wealthy and ultra-wealthy humans) and the divide, or metabolic rift (Foster, 1999), between the city and its hinterlands has deepened. Urban life has increasingly become oriented around work (Graeber, 2018), which has done little to ease the disconnection between urban humans and the land. The rise of the backyard chicken reveals a desire to reconnect outside of the city and to disrupt these urban histories. As one urban smallholder, Theresa, told me, keeping chickens and growing vegetables builds a 'bridge between the country and city':

> A lot of people living in the city yearn for the country, for the open space. It's all very well to yearn for the country and blue skies and rolling hills, and everyone thinks they would love to wake up to the sound of chickens, but you would not want to wake up to the sound of cockerels! If we're not careful, people end up despising and resenting where they live. Being on a crowded Tube or living in a tower block or commuting into work, it can have lots of disadvantages, but it's great to be in a bustling place where you can walk anywhere and there's enough shops and pubs and restaurants. It's trying to get the balance of both. A big part of what we're doing is living the country life in the city, looking at all the things we love in the country and working out how we can still enjoy and benefit from them.

Framing chicken keeping as an urban intervention transgresses the border of the garden, not just bringing a little bit of nature into the city for one residence but challenging the idea of the garden as a controlled and cultivated space. Trevor (2013, p. 2) suggests:

there is still a belief, held amongst city dwellers, that the disorderly forces of nature can be kept at bay; as if the continual, relentless process of growth and decay can somehow be halted or contained by the imposition of human order and stability. Foreign invaders must be expelled from the garden; nettles cut back, thistles up-rooted, dandelions stamped out.

The urban chicken, brought in as a kind of symbol of a simpler life, challenges this orderly and contained vision of urban nature. Gardens don't just 'fill gaps' in the city (Stoetzer, 2018); they accommodate unintended ecologies within and beyond them. Urban flocks are not contained to a single household but connect with multispecies worlds throughout the local area either through interactions with humans or other animals or shifting the makeup of biodiversity. These seemingly ordinary birds are thus creating pockets of multispecies community-building in the city, bringing these alternative urban imaginaries to life. In the next section, I briefly address the role played by class and gender in shaping the chicken city and its urban imaginaries.

Class, gender, and city chickens

Backyard chicken keeping has been framed as a means for middle-class urban residents to live more 'eco-conscious' lifestyles (Barr, 2012) and, more recently, as a protectionist response to the threat of food shortages in English supermarkets (Kearley, 2020). The 'do-it-yourself' surge during the pandemic has been led by 'hipsteaders' with 'a new attitude about what's cool and … the pandemic is accelerating it. Like hipsters, they're setting new trends and flaunting the look on Instagram. But they're also doing a lot of the hard, survival-focused work that defines homesteading' (Kary, 2020).

Urban chicken keeping is not usually undertaken by people relying only on the birds for food; indeed, most chicken keepers found the time and financial costs involved in keeping chickens to be surprisingly high. Although the basics of food, shelter, and bedding can be purchased cheaply, many urban hens are fed specialist feed, housed in premium coops, and offered treats and toys for enrichment. In part, this is because chickens are viewed by many as 'pets with benefits' (Gaffikin, 2009). Being able to afford to keep chickens in one of the most expensive cities in the world points towards a distinct amount of financial privilege and security.

The urban middle classes who keep chickens and cultivate these alternative urban imaginaries do not perceive themselves as 'elites', but frame their situations as lucky rather than a result of systemic privileges or class security. As Tom told me:

> We've been lucky – [our house is] not huge, it's relatively small but it's quite big for London. Probably 7 or 8 meters, a little square. We live in a 1930s semi with a little patio and a garden.

According to Lawler (2005, p. 249), a 'normative and normalized middle-class location is given added legitimacy by a perceived decline in the significance of class itself'. Lawler (2005, p. 436) continues that these narratives of the decline of class importance are 'curiously ahistorical'. Urban chicken keeping's perception as a kind of 'return' to a simpler life connects today's middle class with a romanticized view of the past. For Joe, this was not just to a different time but also a different place:

> We started keeping [chickens] in March 2020. It seems like we made a decision based on lockdown, which is not the case. We'd been thinking about getting them for a while. My wife grew up in Trinidad, where they had lots of chickens, so it was something that she always wanted to do.

In addition to the classed nature of domestic urban chicken keeping, there are also (interspecies) gendered experiences playing out in the chicken city. Traditionally, chicken keeping has been 'women's work', but this is not a stable vision. In the late nineteenth century, domestic chickens were kept in large numbers in London. The 1869 *Cassells Household Guide*, a book designed to include 'everything that a housewife needed to know to run an efficient household', contained a long section on how to successfully keep, feed, and cook chickens. In the early twentieth century, British cities were sanitized of animals, with food production being scaled up and exported (Otter, 2020).

However, during and after the two World Wars, when Britain's food chain was in jeopardy, the number of chickens once again surged in the city, particularly in working men's allotments (Acton, 2011). Of course, chickens never left the city, but there was a distinct rise and fall in their popularity over time and in *who* was taking care of them. As geographer Alice Hovorka (2012, p. 875) writes on chickens in Botswana, 'chickens garner much less attention [than cattle], wield little status and power, and feature in low-valued domestic subsistence or impersonal industrial agriculture realms'. While working with cattle reflects a high social status, caring for chickens is seen as simpler, lower-valued, and individualized work. These attitudes are reflected in recent British agricultural history, where women's labour on family farms has largely been invisibilized (Sayer, 2013). While there is little published research on gender and contemporary urban chicken keeping, Blecha and Leitner (2014) suggest that there are gender differences in chicken care, with women more likely to seek veterinary care. Karabozhilova et al. (2012) received twice as many female respondents to their survey on chicken keeping in London than male respondents – 20 females compared with 10 men.

Chickens' long entwined history with humans has long made them linguistic reference points (Smith and Daniel, 1975), many of which have heavily gendered implications: to be henpecked (nagged by one's wife); cocksure (male arrogance); a mother hen (overprotective); and hen party, for example. Chickens – both hen and cockerel – are memorialized in language. These preconceptions – of hens as overprotective and gentle and cockerels as aggressive and domineering – have material consequences for chickens. Cockerels are not allowed to live in the city limits, and most keepers would not consider keeping a cockerel anyway. For hobby breeders, the 'cockerel problem' means male birds will almost always be killed.

In the urban backyard, different social ideologies, and relationships, are written onto and disturbed by the presence of chickens. In the next section, I explore how these urban interspecies sociabilities play out.

Urban interspecies sociabilities

In the chicken city, new sociabilities flourish between humans, chickens, and other species. This urban milieu troubles categorizations of predatory/prey, pest/pet, and eater/eaten (Taylor and Signal, 2009), even as the consumption of these very same birds continues to increase. Chicken is the most popular meat in the United Kingdom, and is continuing to grow (Eating Better, 2020). People are choosing chicken as they believe it is better for both their health and the environment (Eating Better, 2020), especially in the wake of widespread consciousness-raising campaigns around cow farming and methane emissions (McGregor and Houston, 2018). This is despite broiler chickens literally reconfiguring the Earth's biosphere (Bennett et al., 2018, p. 7) as 'a distinctive new morphotype with a relatively wide body shape, a low centre of gravity and multiple osteo-pathologies'. They are 'a marker species of the proposed Anthropocene Epoch' (Bennett et al., 2018, p. 2) and 'the billions of chickens eaten each year have … fossilized remains [that] will be found as a different species to *Gallus domesticus*, the latest evidence of a human-altered nature in a 10,000-year shared history' (Oliver, 2022).

In London, fried chicken shops are a cultural institution (Hatchman, 2020). Thompson et al. (2018, p. 7) argue that 'the "chicken shop" is a ubiquitous feature of disadvantaged urban neighbourhoods' in East London. Interviewing residents, they found that while chicken shops were the 'defining feature of an unhealthy food environment' (2018, p. 9), they were also culturally valued environments. Against this backdrop, the rise of the urban backyard chicken is even more jarring to larger narratives of London and its collective sense of self. As Amin and Thrift (2017, p. 18) contend, cities 'rely on organized forms of cruelty to non-humans to maintain their human momentum: cities are hungry predators on other forms of life … cities have nearly always been built on the cries and screams and howls of dying animals'. Typically, chickens enter the city as food (Oliver, 2021b). Bringing chickens – alive and lively – into the city's borders might reproduce, reduce, or transform the cruelty between humans and chickens.

In the following section, I explore the social transformations extending from urban chicken keeping in three ways: (1) between humans and chickens in the domestic space; (2) through the impact of chickens on the wider urban multispecies community; and (3) by examining how chickens can expand urban communities.

Human–chicken sociabilities

> When we first got chickens, we were both vegetarian, but we still used to eat their eggs. That was the last thing to go because – and I still believe this – the least harm is eating eggs from your own backyard hens. But it became a philosophical thing: it just felt wrong. I think there's not many people in the world who can understand, but my interaction with those chickens is not transactional. Even if I did want to eat the eggs, I wouldn't.
>
> (Rich)

Before he rehomed chickens, Rich ate animals. In fact, he told me that on the way to collect his first chickens, he stopped to eat a chicken burger: 'Don't hate me for it!' After rehoming chickens, Rich turned vegetarian and, soon thereafter, vegan. In the process of caring for chickens, he connected with them in a way he had never imagined. When chickens are enclosed in farms, the production of food animals is rifted away from consumers. On this scale, chickens morph from an individual being into an absent referent (Adams, 1990) – a *thing*, separated from the human or even the animal world. This view of chickens is challenged when people live with them, leading to advocacy for the species, even if people don't necessarily stop eating them.

Anne runs a chicken business that rents hens out for educational purposes. She offers a range of courses in hatching, rearing, and keeping hens to schools, prisons, and mental health facilities and was effusive about the benefits everyone could gain from living in close proximity to chickens:

> Compared to traditionally what you might get in a school classroom – a hamster or a guinea pig or a rabbit – chickens are very interactive. It's why humans domesticated them in the first place! They are naturally nosy, inquisitive creatures. We do a project called 'the library chicks' and get children to read to them. Especially for children that don't feel confident reading out loud, they'll happily sit on a beanbag and read to the chickens. The chickens will stand and watch them. They might even get a bit of verbal feedback. For some children, it's the only reason they go to school.

Anne began keeping chickens 17 years ago and, after getting over her initial phobia of birds, decided to turn her hobby into a business with a social conscience. A local school got wind of her chicken keeping and approached Anne to help them establish a coop. Anne found someone to build them a coop and got six ex-commercial hens from the British Hen Welfare Trust (then named the Battery Hen Welfare Trust). The school was a 'pupil referral unit', a type of school in Britain for children unable to be in mainstream education – usually because they have been excluded. When Anne brought the hens in, the teacher said they'd be happy if the students engaged for even 10 minutes:

> Forty-five minutes later, I have six teenage boys sitting in the coop with chickens on their lap. We were talking about dinosaurs and where chickens came from. We could have stayed there all afternoon. I thought, if this is the impact six scrawny ex-commercial girls can have on teenage boys who can't cope in mainstream school, what impact can they have across all settings?

Different kinds of 'chicken projects' are introduced for many reasons in schools: for mental health, learning about animals, and connecting chickens to the curriculum. In urban schools, this is often a rare opportunity to connect with 'food' animals. Acknowledging the many benefits of hens, the British Hen Welfare Trust is currently researching and developing a 'Hens as Therapy' programme to aid people with mental illness. This kind of more-than-human therapeutic landscape (Gorman, 2017) can have meaningful effects and create a sense of belonging for patients (Gorman, 2019). However, there is often little consideration of the birds themselves. For urban keepers, there have been positive effects from caring for chickens in terms of their relationships with them:

> They're quite underestimated creatures and I'd like more people to appreciate them as pets. We have so much language that comes from chickens. Every time I say something like 'being cooped up' ... so much of what we say comes from chickens, but so many people don't really have much experience of being around them or much knowledge of them as a pet that lives outside and pumps out food for you regularly. They're good for your mental health.

Urban mental health has crystallized as a critical concern for both the present and the future. Anarchists such as David Graeber (2018) point to the specific violence of urban living and work that eat into social life: commuting hours across the city; having little time for family; living in high-density areas; and a lack of access to nature or more-than-human communities – especially for renters. Many of the people I interviewed were largely buffered from the contemporary urban violence of, for example, housing dispossession (Pain, 2020) or gentrification and displacement (Atkinson, 2000). Nonetheless, the pandemic has presented well-being and health-related challenges.

The supposed ills of urban life can lead to the middle classes desiring to either leave or transform the city, but for some urban keepers, keeping chickens counteracted this desire by creating different, yet still instrumentalizing, encounters with nature in the city. At the end of 2020, there was an avian influenza outbreak in Britain that overlapped with national restrictions in England due to COVID-19. Theresa followed the avian lockdown regulations closely:

> I personally haven't found lockdown difficult but that's because I work from home. I know that the rest of my family have found it more challenging but then with the avian flu lockdown, it's like you're locked down, we're locked down, we're all locked down together, you know, welcome to lockdown!

While Theresa, who had kept birds for many years before the lockdown, saw little change in her relationships with the chickens, she did recognize their shared experiences of lockdown: of boredom, containment, and enclosure. For Joe, who took in his first chickens in March 2020 at the beginning of the pandemic restrictions in England, this timing offered a relationship with the hens that he might otherwise not have had:

> We've had an artificial situation in lockdown. For a lot of time, it's been just us and the chickens, so we've noticed their personalities more than we would have done otherwise. We've been surprised that they're wonderful pets. They've got certain reflexes and behaviours that they can't control. If you put them somewhere where you don't want them to dig up the plants, they will dig up the plants, because that's what they do, but they do seem to learn.

Lockdown offered many keepers more time with their chickens, or even opportunities to keep chickens for the first time. Reflections on keeping birds seemed to be lengthy and inflected by this shared state of lockdown. In the urban barnyard, new sociabilities emerged between humans and chickens, but these were rarely liberatory, instead reproducing unequal dynamics. In the next section, I consider how these sociabilities affected much larger multispecies communities in the city.

Multispecies sociabilities: Predation, biodiversity, and biosecurity in the galline garden

Animals have always been urban actors, but their stories have often been subsumed into human narratives (Van Dooren and Rose, 2012). The non-human storying of the multispecies city (Van Dooren and Rose, 2012) can provide new perspectives on the world and transform relationships with other animals on both the individual and species levels. In the galline urban garden, a whole other beyond-human world is opened up, partly through the violence of urban predators – and the biggest perceived threat to urban chickens is their city rival: the fox:

We've had foxes that have dug in and gone in underneath the coop. We learned that the coop has to be surrounded so they can't dig in, or wire all the way around, or have a solid base … that's fantastic Mr Fox, he is a fantastic fox! They are programmed to get free food.[4] I do my absolute utmost, and my daughter spent money trying to protect the girls, but he was clever enough to strike that one night that we forget to put them away properly. Bad luck to the chickens – that's sad, but Mr Fox might go and feed his new cubs.

(Jenny)

A citizen science survey recently found evidence that the number of foxes in English cities has doubled since the 1980s (Scott et al., 2018). Urban foxes are a common phenomenon across the world, and there is some evidence that living in the city has led to 'phenotypic divergence' in skull morphology between urban and rural foxes (Parsons et al., 2020). Studies of urban foxes – and human attitudes towards them – have been characterized into positive and negative interactions (Morgan and Cole, 2011; Soulsbury and White, 2019), with the most serious of the latter including killing pets. However, as Jenny describes, the urban fox isn't necessarily seen as an enemy to these urban chicken keepers. There is an empathetic extension to the predators that threaten their flocks, with the blame often being shouldered by human error in caring for the hens. Urban chicken keepers are aware of their enclosed chickens being an appetizing – and potentially easy – meal for local foxes and adapt their protective measures accordingly:

At the weekends and during the day while we're at work, they will stay in their run. We don't trust foxes and everything else around here.

(Saffron)

Several of the people I interviewed had experienced fox attacks, with some losing their entire flocks. Keepers often installed infrastructure to protect their chickens in anticipation of local foxes:

We have a little eglu [plastic coop] with an eight-foot run coming off it. For a while, we had a flexible nylon fence that we could move around the garden with the idea that they could free range a bit, but now we've got a completely enclosed walk around because we have foxes here. We've had a couple of fox attacks, we haven't lost any chickens to a fox, but we have had some fox experiences so now they're in what I refer to as chicken prison but we're going to get some plants to make it look a bit nicer.

(Lisa)

The galline garden is deliberately designed in conjunction with the perceived threat of foxes. Backyard chicken keeping does not exist in a vacuum, nor is the galline garden a closed ecosystem: just as foxes can creep in, so can chickens draw attention to the other lives in the garden. Jenny, for example, has noticed that while she doesn't have many wild birds in her garden (probably due to her cats), there are pigeons and 'very occasionally the pigeons will actually brave going into the run and tackle the chickens, but it's actually the chickens who will attack the pigeons and just go get out this is ours'. Marie has lots of wild birds in her garden, which she feeds, but they rarely interact with her chickens. However, her neighbour's cats have been drawn to the chicken coop:

> There is a black cat that looks quite young and it sort of stalks them down the side of the garden, and I think the chickens know that they're secure in there, because they don't seem to move away from where the cat is. They just kind of stand and look at it and it's kept sort of crouched down as if it's doing this amazing stealth thing down in the grass.[5]

As well as cats and wild birds, chickens are themselves predators of other forms of life: worms, slugs, snails and even plant life. The galline garden shifts relations and ecologies, producing new relationships of not just predation, but of fascination, play, and ecological changes.

Chickens have many interactions with other animals in the garden, which can also mean risks of disease outbreaks. During this research period, poultry birds in Britain had been placed in lockdown due to a highly pathogenic avian influenza (H5N8). For backyard keepers, the main understanding of this measure was that they needed to keep their flocks away from migratory water birds, which were most likely to carry the disease:

> I have read [DEFRA guidelines] and when there's an alert goes out, like the bird flu, I get it. I understand where that's coming from if one were doing it commercially or if I was selling the eggs, but I can choose to ignore it. I completely understand how dangerous that is, but it's a bit like following the rules for the pandemic really. We have our birds, but we don't work, so we're home, and I've never seen ducks in the garden, and I've never seen wild fowl. I have currently got three cats and they all have bells, but they are hunters who catch wildlife, lots of mice but not so many birds. I don't have many wild birds in our garden, apart from pigeons.
>
> (Jenny)

> I have a swan's nest in my garden; we have two swans. In my garden, there's swans, there's chickens, there's cats, they're all knocking along nicely together ... I reduce the risk of disease by making sure [the chickens] are kept with good fresh litter in their house, nice and dry. When I clean them out, I put diatomaceous earth in, I spray for mites. It is a natural product, so I spray the birds sometimes too. I clean their dirty bottoms as well to stop worms going up them.
>
> (Elizabeth)

Jenny and Elizabeth know one another through the online chicken-keeping community, but they don't live close to each other in London. They also have very different perspectives on the risks and responsibilities in the multispecies garden and their reflections on 'good' and 'bad' animals in the urban garden. While Jenny chooses to ignore the bird flu regulations based on her own knowledge of her garden, the local area, and the non-human populations, Elizabeth is incredibly dedicated to keeping the birds she knows are nesting in her garden, as well as her chickens, safe. While only two examples of a range of opinions about non-human animals, it is clear that keeping chickens can shift the perceptions that chicken keepers have of other species. New 'necessary evils' emerge in the keeping of chickens: spraying for mites, eradicating pests, and keeping out predators. For most people I interviewed, predators (aside from rats) weren't positioned as inherently 'bad', but were in need of 'controlling'. When telling me about her looser interpretation of the avian influenza lockdown rules, Jenny compared her relatively wildlife-free (or at least of wild fowl) garden to Elizabeth's garden, where swans nest, indicating that risk reduction is situational and changing:

> Elizabeth is on a river and has swans and ducks and everything else in her garden. She's got swans and ducks and geese sometimes come along and it's such a transient community, they do move around. I understand why she would be far more inclined to keep her birds completely safe and indoors.

Elizabeth's swans began nesting in her garden, which backs onto a river, four years ago. At the time, she was already a well-established urban chicken keeper, having kept a small flock for about 10 years. While Elizabeth didn't see a connection between her chickens and the swans nesting in the garden, she did speculate that perhaps her garden felt like a safe haven for wildlife. However, with the avian influenza outbreak, this multispecies urban encounter between humans, swans, and chickens was inflected with potential danger.

When people decide to take up urban chicken keeping, they rarely anticipate the more-than-human communities – and responsibilities – they are entering. Novelist Alice Walker (2011) says keeping chickens opens a parallel world in which all other animals exist, and more recently the love and strife of keeping chickens has been fictionalized by Jackie Polzin in her novel *Brood* (2021). This certainly seems to be true for these chicken keepers, whose gaze and interest have been drawn down to the ground level. Where previously the garden was interacted with from above, landscaped to accommodate human residents, and their social visions and political ideals (Weiner, 2003), the chickens offer new perspectives and sociabilities with and beyond the galline flock. In some cases, it can even be an emancipatory project: what Cudworth and Hobden (2017, p. 35) see as part of a project that can 'explore, experiment, and consider futures that are alternate to the current neoliberal path'. But on that path remain cruelty and changing priorities with regard to other species:

> One of the joys is giving the girl snails. When one of them gets a snail in their beak and they wonder if there's one there, and they make this noise – they make so much noise, then they drop [the snail] and then, one of the others grabs it. It's just been hilarious.
>
> (Jenny)

In the city, chickens are not only producers for humans nor only prey for foxes or predators of small invertebrate animals. These categorizations of chickens can oversimplify complex urban multispecies communities and erase the agency and lives of individual chickens (Barua and Sinha, 2019), each of whom has its own unique personality, relationship, and connection with the urban space. While they are enclosed in gardens and backyards (even more heavily during avian lockdowns), chickens are not separate from the more-than-human city, but overflow and transgress the spaces assigned to them. They interact and build relationships – both positive and negative – with other urban animals, which comprise valuable communities in and of themselves.

Watching and learning about the wider multispecies urban community – and their relationships with the chickens, including the threats posed to the flock – curates a different relationship for humans with the city. In the next section, I explore how urban flocks can expand *human* and *non-human* communities.

Expanding community: Interacting with urban chickens

In Britain, 'livestock' animals are regulated by two government bodies: the Department for the Environment, Food, and Rural Affairs (DEFRA) and the Animal and Plant Health Agency (APHA). Backyard or domestic chickens, while subject to regulations and controls, are not required to be registered with either agency if the flock is smaller than 50 birds. While hen-rehoming organizations encourage keepers to register their birds, almost no one I interviewed had done so. Local councils or house deeds can prohibit the keeping of backyard chickens (or specifically cockerels) for a variety of factors pertaining to noise or hygiene. In this section, I explore how bringing chickens into the city can change local relationships with neighbours and strangers.

When I began interviewing backyard keepers, I expected to find that keeping chickens would cause antagonistic relationships with the local community, a concern echoed by Seb:

> In 2019, myself and my partner were looking to buy a house here [in London], but at that time we knew that we were looking to buy a house in this area that had to have a garden and it had to be suitable to have ducks and chickens in within our budget, because obviously the best

properties for that are like £900,000 – well above our first house! We wanted space to breed the birds but to manage it in a way that wasn't a nuisance to our neighbours.

While backyard keepers were conscious of the fact that they lived 'all on top of each other here' (Saffron), very few had their fears realized. One keeper, Jenny, didn't have a noise complaint, but her hens did attract rats to her garden, which subsequently led to rats being found in her neighbour's child's paddling pool. Rats are often attracted to chicken coops as food can be left out or spilt, making these easy pickings for rats. While rats don't usually attack chickens (ChickenGuard, 2020a), the allure of feed and especially freshly laid eggs can cause them to nest near small domestic flocks. According to one advice site, 'rats carry diseases that are harmful to chickens, will attack baby chicks, steal eggs, and have even been known to chew on hens' feet while they are sleeping' (ChickenGuard, 2020b). They are also unwelcome neighbours to the humans in the area. The species that come along with chickens can often prove to cause more tension than the chickens themselves.

Despite this 'dirty' side of chicken keeping, urban keepers were more focused on the value and joy that chickens brought to them and their local community. Jenny told me that, despite tensions over rats, people loved to interact with her chickens; she runs a business from home and found that:

> I get [customers] come and lots of people with younger children. It's lovely to introduce children to the chicken and ask: 'Would you like to see if they've made an egg today?' The neighbours who got the rats have youngsters and quite often, one of them will come over to the fence. Once she saw [the chickens] and that we had eggs. She cradled it and I said, 'You can have that for lunch today,' so she did. Her mum said she's never eaten eggs before! Now, she loves eggs. Quite often, my neighbours get eggs. I love seeing little ones in absolute fascination by it, realizing where eggs come from.

The connection between chickens, education, and the wider local community was an especially strong theme for many backyard chicken keepers, but this came with concerns about the socialization of children into utility relationships with these birds. Talking to two different 'hen professionals', I raised this question. At a non-slaughter city farm in London, the hen keeper told me that introducing children to chickens in their education has a range of purposes, but their ultimate goal is to teach children about the *responsibilities* we have to animals, regardless of whether we eat them. At a non-slaughter city farm, chickens have far more to offer than just eggs – cuddles, therapy, and education – but the very existence of these spaces places animals in a position of duty to humans. There was a sense that children growing up in urban environments were missing out on learning about animals, nature, and where their food came from, and that this was vital to help them understand how to treat animals well and with respect. Anne, who runs a hen-hire business, shares this opinion:

> We show [the children] how to handle the hens. They've got lots of check-lists on the coop to show them how to look after them, designed for children. For a lot of children, it's their first interaction with an animal, so there's the responsibility of caring for that animal. Chickens are brilliant because if they're not being looked after properly, they will express that opinion quite loudly. Children generally have respect for the chickens from the get-go.

Whether through chicken programmes in city schools or backyard hens inviting conversations with neighbours, when hens move into a neighbourhood, far more changes than can be confined to a single backyard. Neighbours get involved over fences or in consuming the eggs of these city birds; keepers and their friends report improving not only their well-being in the city but also their connection with the world beyond the human. In many cases, the presence of chickens changes people's opinions of these birds, inducing more respect and fascination with the chicken. Even if people don't stop eating chickens, it can create space for conversations around chicken liberation. One interviewee said:

> A lot of our friends like coming over. They bring their kids and, once a year, we try and throw a little party for the anniversary of 'being out of the cage'. Partly, it's a subtle animal rights thing, to show people this is what chickens are like. It's also just nice to get people over for a party where the chickens are out, lots of lovely vegan food, and meet the chickens. During COVID, it's nice to just sit there with a cup of tea and chickens clucking around. We live in a busy part of Islington and it's like an oasis with a different filter.

This, like the other examples in this chapter, shows how urban chickens don't just form relationships with the humans with whom they live, but with the wider social and local networks of their humans – from friends and family to neighbours to random visitors like customers or postal workers. This can be put to work in different ways: socializing into animal use or socializing into animal liberation. The cultural reproduction of animal harm is found throughout childhood, from education to media to family practices (Cole and Stewart, 2016). As Lynda Korimboccus (2021) argues in 'The Peppa Pig Paradox', the agents socializing children (media, education, healthcare, and governments) reproduce anthropocentric and speciesist ideals that categorize some animals as 'food'. Urban flocks challenge these categorizations: of who we eat, of what animals belong in cities, and even of whom we should interact with in close contact as a species. In the conclusion to this chapter, I discuss how these new urban interspecies sociabilities are aiming to usher in a 'new normal' for chickens; crucially, though, this is against a backdrop of increasing exploitation and continued instrumentalization of these birds.

The chicken city

Throughout this chapter, I have focused on interspecies social relationships that are emerging between chickens, humans, and other animals in London. In these chosen chicken–human spaces, social lives emerge not only between chickens and humans, but also with the flora, fauna, and wider more-than-human communities of the city. This multispecies urban milieu troubles categorizations of predatory/prey, pest/pet, and eater/eaten, even as the urban population at large continues to rely on the churn of consumption of these very same birds. Keeping chickens is not simply a beneficent or non-violent relationship – indeed, for some chickens, the urban backyard is not a haven of safety but a space of enduring violence. A pandemic surge in chicken keeping has led to some changing their minds about chickens, but nowhere near on a revolutionary scale. For most domestic chicken keepers, the back garden might be an alteration in the relationship between humans and chickens, but it is far from a transformation. Even with the knowledge and relationships formed, for some humans, chickens remain firmly outside of circles of care.

While it might be easy to assume that chickens would only affect the humans with whom they are living or the creatures with whom they share the garden, bringing chickens into the city can create all kinds of new social relationships. Increasingly, knowing and caring for chickens is being perceived as positive for human well-being, leading to more chickens living in collective spaces such as schools. Urban keepers and carers attested to the amazing benefits that chickens had brought to their lives, not only through their eggs but through spending time with them. Most – if not all – of the people I interviewed spent a long time researching the behaviours of chickens and how best to provide a safe and nourishing environment for them that is in contention with the usual (non-) place of chickens in the city. However, these new sociabilities are not neutral; in the urban barnyard and garden, chickens are still under the control and care of humans and the welfare of chickens is always secondary to that of the humans. The chicken city is not a liberated space devoid of violence; rather, as it grows, it presents opportunities for understanding how humans are choosing to live differently with animals, and how this transforms the perceptions of 'farmed' animals and the city itself.

Notes

1 All names of interviewees and their businesses/organizations are pseudonyms.
2 Chickens' names are also pseudonymized.
3 This chicken-keeper bred and reared show chickens but only hens, not cockerels, are 'shown'. However, chickens can't be sexed without technology until they are six to eight weeks old, and cockerels cannot be kept in built-up areas. Half of all chicks are male. In commercial agriculture, these are sexed and culled at days old, or even in vitro (Krautwald Junghanns et al., 2018). For hobby breeders, the 'cockerel

problem' is solved by killing (and often eating) young birds as soon as they can be sexed.

4 The idea of food as 'transactional' for wild animals is an interesting one: what, exactly, wouldn't be 'free food' for a fox? In his novel *Fox 8*, George Saunders (2018) writes as a suburban fox – including a foxy kind of spelling – whose habitat is over-taken by a new shopping mall, disappearing the rest of his pack. Fox 8, the narrator, overhears a 'fawlse and meen' story a 'yuman' is telling their child: 'In that story was a Fox. But guess what the fox was? Sly! He trikked a chiken! He led this plump Chiken away from its henhowse … We do not trik Chickens! We are very open and honest with Chickens!' The deal, the fox writes, is that the chicken lays eggs and the fox eats them. If a chicken doesn't run, they consent to be eaten. Jenny's 'free food' comment writes human/non-human relations onto a very different kind of non-human relationship that can't be understood through capitalist transactional relations.

5 Noting the difference between 'they' and 'it' in referring to the animals that Marie keeps and the predator she is trying to keep away positions an interesting linguistic binary between her hens and another cat, despite the hens having little interest in or feeling threatened by the cat.

References

Acton, L. (2011). Allotment gardens: A reflection of history, heritage, community and self. *Papers from the Institute of Archaeology*, 21, 46–58.

Adams, C. (1990). *The Sexual Politics of Meat*. Cambridge: Polity Press.

Amin, A. and Thrift, N. (2017). *Seeing Like a City*. Cambridge: Polity Press.

APHA. (2019). *Common Diseases in Backyard Poultry Flocks in Great Britain*. http://apha.defra.gov.uk/documents/surveillance/diseases/leaflet-backyard-poultryguidance.pdf

Atkinson, R. (2000). The hidden costs of gentrification: Displacement in central London. *Journal of Housing and the Built Environment*, 15(4), 307–26.

Atkinson, R. (2020). *Alpha City*. London: Verso.

Barnhill, A. (2015). Does locavorism keep it simple? In A. Chignell, T. Cuneo and M.C. Halteman (eds), *Philosophy Comes to Dinner*. New York: Routledge.

Barr, D. (2012, 21 March). The eggs factor: How the middle class fell in love with chickens. *The Telegraph*. www.thetimes.co.uk/article/the-eggs-factor-how-the-middle-class-fell-in-love-with-chickens-vjl0mtsvfrd

Barua, M. (2017). Non-human labour, encounter value, spectacular accumulation: The geographies of a lively commodity. *Transactions of the Institute of British Geographers*, 42(2), 274–88.

Barua, M. and Sinha, A. (2019). Animating the urban: An ethological and geographical conversation. *Social & Cultural Geography*, 20(8), 1160–80.

Bennett, C.E., Thomas, R., Williams, M., et al. (2018). The broiler chicken as a signal of a human reconfigured biosphere. *Royal Society Open Science*, 5(12), 180325.

Blecha, J. and Leitner, H. (2014). Reimagining the food system, the economy, and urban life: New urban chicken-keepers in US cities. *Urban Geography*, 35(1), 86–108.

Chappell, B. (2020). 'We are swamped': Coronavirus propels interest in raising backyard chickens for eggs, *NPR*. www.npr.org/2020/04/03/826925180/we-are-swamped-coronavirus-propels-interest-in-raising-backyard-chickens-for-egg?t=1633445601137

ChickenGuard. (2020a). Do chickens attract rats? www.chickenguard.co.uk/do-chickens-attract-rats

ChickenGuard. (2020b). The ultimate guide to keeping rats away from your chicken coop. www.chickenguard.us/the-ultimate-guide-to-keeping-rats-away-from-your-chicken-coop

Cole, M. and Stewart, K. (2016). *Our Children and Other Animals: The Cultural Construction of Human–Animal Relations in Childhood*. London: Routledge.

Collard, R.C. and Dempsey, J. (2013). Life for sale? The politics of lively commodities. *Environment and Planning A*, 45(11), 2682–99.

Cudworth, E. and Hobden, S. (2017). *The Emancipatory Project of Posthumanism*. London: Routledge.

DEFRA. (2018). Poultry registration rules and forms. www.gov.uk/government/publications/poultry-including-game-birds-registration-rules-and-forms

Eating Better. (2020). *We need to talk about chicken*. www.eating-better.org/uploads/Documents/2020/EB_WeNeedToTalkAboutChicken_Feb20_A4_Final.pdf

Foster, J.B. (1999). Marx's theory of metabolic rift: Classical foundations for environmental sociology. *American Journal of Sociology*, 105(2), 366–405.

Gaffikin, B. (2009, 26 February). The new urban hens are often pets with benefits. *Civil Eats*. https://civileats.com/2009/02/26/the-new-urban-hens-are-often-pets-with-benefits

Gaffney, A. and Kinninment, M. (2020, 19 March). Backyard chickens in huge demand as coronavirus shopping frenzy empties supermarkets. *ABC News*. www.abc.net.au/news/2020-03-19/backyard-chickens-in-demand-due-to-coronavirus/12068392

Gorman, R. (2017). Smelling therapeutic landscapes: Embodied encounters within spaces of care farming. *Health & Place*, 47, 22–8.

Gorman, R. (2019). Thinking critically about health and human–animal relations: Therapeutic affect within spaces of care farming. *Social Science & Medicine*, 231, 6–12.

Graeber, D. (2018). *Bullshit Jobs: A Theory*. New York: Simon & Schuster.

Griffith, G. (2020). Coronavirus: Lockdown rush for hens prompts supply concern. *BBC News*. www.bbc.co.uk/news/uk-wales-52732470

Hatchman, J. (2020). The best fried chicken in London. *Eater London*. https://london.eater.com/maps/best-fried-chicken-london

Hovorka, A.J. (2012). Women/chickens vs. men/cattle: Insights on gender–species intersectionality. *Geoforum*, 43(4), 875–84.

Karabozhilova, I., Wieland, B., Alonso, S., Salonen, L., and Häsler, B. (2012). Backyard chicken keeping in the Greater London Urban Area: Welfare status, biosecurity and disease control issues. *British Poultry Science*, 53(4), 421–30.

Kary, T. (2020). COVID is turning us all into hipsteaders. www.bloomberg.com/news/features/2020-09-08/covid-diy-boom-raising-chickens-gardening

Kearley, S. (2020). How COVID-19 has affected chicken keepers in UK. *Backyard Poultry*. https://backyardpoultry.iamcountryside.com/chickens-101/covid-19-chicken-uk

Korimboccus, L.M. (2021). Pig-ignorant: The Peppa Pig paradox. Investigating contradictory consumption in childhood. *Journal for Critical Animal Studies*, 17(5), 3–13.

Krautwald Junghanns, M.E., Cramer, K., Fischer, B., et al. (2018). Current approaches to avoid the culling of day-old male chicks in the layer industry, with special reference to spectroscopic methods. *Poultry Science*, 97(3), 749–57.

Lawler, S. (2005). Disgusted subjects: The making of middle-class identities. *The Sociological Review*, 53(3), 429–46.

McGregor, A. and Houston, D. (2018). Cattle in the Anthropocene: Four propositions. *Transactions of the Institute of British Geographers*, 43(1), 3–16.

Mellen, S. (2020, 20 August). Chicken rehoming charity gets 52,000 lockdown hen requests. *BBC News*. www.bbc.co.uk/news/uk-england-53832858

Microsanctuary Resource Centre. (2021). *What is a microsanctuary?* https://microsanctuary.org/what-is-a-microsanctuary

Morgan, K. and Cole, M. (2011). The discursive representation of non-human animals in a culture of denial. In B. Carter and N. Charles (eds), *Human and Other Animals*. London: Palgrave Macmillan.

National Farmers' Union. (2021). Coronavirus: What is the impact on the poultry sector? www.nfuonline.com/sectors/poultry/poultry-news/coronavirus-what-is-the-impact-on-the-poultry-sec/#What%20should%20I%20do%20if%20 I%E2%80%99m%20worried%20about%20looking%20after%20my%20birds%20 if%20I%20fall%20unwell?

Oliver, C. (2021a). Returning to 'the good life'? Chickens and chicken-keeping during COVID-19 in Britain. *Animal Studies Journal*, 10(1), 114–39.

Oliver, C. (2021b). Eating chickens, consuming cities: Urban metabolism in animal studies. *The Animal Turn*. www.theanimalturnpodcast.com/post/eating-chickens-consuming-cities-urban-metabolism-in-animal-studies

Oliver, C. (2022, 8 March). Rising with the rooster: How urban chickens are relaxing the pace of life. *The Sociological Review Magazine*. https://doi.org/10.51428/tsr. hjbn7857

Otter, C. (2020). *Diet for a Large Planet*. Chicago: University of Chicago Press.

Pain, R. (2020). Geotrauma: Violence, place and repossession. *Progress in Human Geography*, 45(5), https://doi.org/10.1177/0309132520943676

Parsons, K.J., Rigg, A., Conith, A.J., et al. (2020). Skull morphology diverges between urban and rural populations of red foxes. *Royal Society Publishing Proceedings B*. https://doi.org/10.1098/rspb.2020.0763

Polzin, J. (2021). *Brood*. London: Penguin Random House.

Saunders, G. (2018). *Fox 8*. London: Bloomsbury.

Sayer, K. (2013). 'His footmarks on her shoulders': The place of women within poultry-keeping in the British countryside c 1880–1980. *Agricultural History Review*, 61(2), 301–29.

Scott, D.M., Baker, R., Charman, N., et al. (2018). A citizen science based survey method for estimating the density of urban carnivores. *PLoS ONE*, 13(5), e0197445. https://doi.org/10.1371/journal.pone.0197445

Smith, P. and Daniel, C. (1975). *The Chicken Book*. Athens, GA: University of Georgia Press.

Soulsbury, C.D. and White, P.C.L. (2019). A framework for assessing and quantifying human–wildlife interactions in urban areas. In B. Frank, J.A. Glickman and S. Marchini (eds), *Human–Wildlife Interactions: Turning Conflict into Coexistence*. Cambridge: Cambridge University Press.

Stoetzer, B. (2018). Ruderal ecologies: Rethinking nature, migration, and the urban landscape in Berlin. *Cultural Anthropology*, 33(2), 295–323.

Taylor, N. and Signal, T.D. (2009). Willingness to pay: Australian consumers and 'on the farm' welfare. *Journal of Applied Animal Welfare Science*, 12(4), 345–59.

The Chicken Vet. (2021). *Vaccinating poultry with the smaller flock in mind*. www. chickenvet.co.uk/news/post/vaccinating-poultry-with-the-smaller-flock-in-mind

Thompson, C., Ponsford, R., Lewis, D., and Cummins, S. (2018). Fast-food, everyday life and health: A qualitative study of 'chicken shops' in East London, *Appetite*, 128, 7–13.

Trevor, T. (2013). Three ecologies. In P. van Cauteren (ed), *Lois Weinberger*. Ghent SHMAK.

Van Dooren, T. and Rose, D.B. (2012). Storied-places in a multispecies city. *Humanimalia*, 3(2), 1–27.

Walker, A. (2011). *The Chicken Chronicles*. London: Phoenix

Weiner, A. (2003). *Landscaping the Human Garden: Twentieth-Century Population Management in a Comparative Framework*. Stanford, CA: Stanford University Press.

2 Fairweather friends?

Rethinking multispecies companionship in the new normal

Zoei Sutton

Introduction

In the throes of the COVID-19 pandemic, lockdowns became an all-too-familiar occurrence. As humans found themselves increasingly restricted to their homes, experiencing limited contact with their own species, many relied on the companionship of non-human animals to fulfil their social and tactile needs for contact. Acquisition of pet commodities skyrocketed, with shelters and for-profit breeders alike reporting that they had run out of animals to meet public demand and pet ownership rates increasing from 61% in 2019 to 69% in 2021 (AMA 2021). These patterns speak to the increasing valuation of the companionship that commodified 'pets' can provide to their human owners. However, there has been significantly less critical discussion on whether humans are good multispecies companions to non-human others (see Sutton 2020; Tuan 1984 for examples of this critique). This chapter will explore this pertinent question by focusing on multispecies relations in the site where many of us spent much of our time during the pandemic: the home.

I begin by exploring sociological perspectives on the use of animals as companions, and the social and physical constructions of 'home' as a human-centric place, setting up the asymmetrical power dynamic that underpins multispecies relations occurring within it. Then, drawing on a review of print media depictions of companion animals during the COVID-19 pandemic, I highlight key themes in how human–pet entanglements were represented in Australia during the global health crisis. From a sociological perspective, media representations play a key role in the development of social imaginaries or authorizing environments for particular norms and values (Anderson 1991). For human–non-human animals, particularly, these representations reinforce anthroparchal norms, bolstering narratives that privilege human wants and needs over those of other animals and the natural environment (e.g. Stewart & Cole 2009). Of particular note is the strong emphasis on the commodification of non-human animals as pet commodities and the navigation of multispecies everyday life that aptly highlights the complexity of these relationships, which are simultaneously built on 'familial' affection and human-centric exploitation. I conclude with a call to rethink the multispecies companionships that

DOI: 10.4324/9781003257912-4

can take place in the home, extending the boundaries of who is a potential companion beyond purchased 'pets' to include the many other creatures sharing home spaces. In doing so, I highlight how human-centrism is currently curtailing the development of rich, multispecies companionship that might be enabled by more inclusive socio-spatial practices, arguing that the pursuit of multispecies companionship is a fruitful avenue through which we might learn to live 'less badly'[1] with others in the new normal, for the benefit of all.

Sociological perspectives on non-human animal companions

While a plethora of sociological literature focuses on the role of companion animals as cherished family members (e.g. Charles, 2014; Cudworth, 2011; Greenebaum, 2004; Irvine and Cilia, 2017), the sociological toolkit has also been harnessed to critically examine the unequal power relations and exploitation evident in these human–animal entanglements. Two key approaches to this critical study are Erika Cudworth's (2011) conceptualization of anthroparchy and Matthew Cole and Kate Stewart's (2014) socio-discursive construction of non-human animals.[2] In this section, I discuss each of these approaches, highlighting how non-human animal companions are situated within them.

The relationship between structure – broader social forces that enable and constrain social actors – and agency – an individual's capacity to act within these structures – has been a key element of sociological approaches to the study of everyday life (Archer, 1995; Giddens, 1984). Cudworth's (2005, 2011) anthroparchy draws on this intellectual history to develop a complex systems approach to dominance. She argues that the term 'anthropocentrism' does not adequately capture the systematic organization of power relations between humans and other species, instead proposing that we conceive of anthroparchy as a dominant system in which human wants and needs are consistently prioritized over those of other animals and the natural environment. This system is characterized by five structures that perpetuate this privileging of the human species: (1) anthroparchal relations in production; (2) anthroparchal reproduction and domestication; (3) anthroparchal politics; (4) violence; and (5) cultures of exclusive humanism (Cudworth, 2011). Anthroparchy intersects with other systems of domination, such as capitalism, patriarchy, and settler colonialism, thus compounding the effects of oppressive structures (see Chapter 7 in this collection). For companion animals, these anthroparchal structures legitimate breeding and domestication processes that see minded beings turned into lively pet commodities – that is, commodities who are alive throughout their commodity life and whose liveliness is a key aspect of their appeal (Collard, 2014). These 'pets' are then situated in a social world that is shaped around exclusive humanism such that their wants and needs are subordinated below those of humans – even in cases where owners may attempt to minimize the power inequality in their multispecies relationships (Cudworth, 2011; Sutton & Taylor, 2022). Any companion animal not meeting the expectations of pet commodities, for instance, in providing unconditional affection and entertainment, can be unmade and relinquished or euthanized when they are no longer wanted – their lives are only valuable when considered so by humans (Collard, 2014; Irvine and Cilia, 2017; Sutton, 2020; Sutton and Taylor, 2022).

This making and unmaking of non-human animals as pet commodities introduces another key approach to critically understand human–pet relations – the socio-discursive construction of non-human animals. Cole and Stewart (2014) argue that non-human animals are socio-discursively positioned along a continuum of objectification, depending on their level of (dis)utility to humans. Non-human animals with whom humans may have affective ties are more likely to have 'sensible' lives – that is, the material conditions in which they live are more likely to be visible or knowable to humans. Those that are violently exploited, such as animals farmed for their flesh, are likely to be non-sensible, their lived realities hidden away (Cole and Stewart, 2014). Importantly, non-human animal roles are not tied to species in Cole and Stewart's conceptualization. Citing the example of a rabbit that may occupy the role of a companion animal, testing subject, farmed animal, 'pest', 'wild', Easter bunny or entertainment, they demonstrate that one species can occupy multiple socio-discursive positions, depending on what humans want to use them for. For companion animals – those constructed as pet commodities – the shared meanings around 'pets' determine how they will be interacted with and how sensible their lives are. Although 'companion animals' are positioned as one of the most subjectified, sensible categories, it must be noted that elements of domestication and breeding are often hidden from view in order to maintain a view of pet ownership that is wholly positive and familial (Sutton, 2020). Further to this, the cultures of exclusive humanism outlined by Cudworth are such that the sensibility of companion animal lives likely represents a human-centric view – one that doesn't seriously consider the experiences of animals used as pet commodities but instead reflects the narratives humans weave around pet ownership. The complexity of socially constructing non-human animals as pets is aptly captured by historian Erica Fudge (2008), who surmises that human–companion animal relationships are enabled by two contradictory discourses: (1) that humans have a natural mastery over non-human animals, and animals' instinct is to submit to human dominance and (2) that non-human animals love humans and want to be owned. The former legitimates the breeding and selling of offsprings, for non-human animals are seen to have no capacity for emotion; this raises the question of why humans would want to own pets if they are unable to love 'their' owners. The latter acknowledges animal companions as minded emotional beings but then raises ethical concerns around the forced separation of these beings from their families in order to fulfil human wants for purchasable companionship.

Anthroparchy in the home

As the above discussion indicates, anthroparchal power relations are evident in the discourses, politics, and practices of everyday life, and the home space is no exception. Henri Lefevbre (1991) argues that space is socially constructed, shaping and being shaped by the social context in which it occurs. For multi-species families, the home is inescapably steeped in asymmetrical power relations that maintain human wants and needs over those of non-human inhabitants (Sutton and Taylor, 2022). Home is an important source of ontological security, and positive associations are challenged when inhabitants cannot live freely or control the space (Douglas, 1991; Fudge, 2008). This can result in humans stifling the agency of their companion animals, which are seen to engage in undesirable uses of the home, such as creating a 'mess', climbing furniture and screen doors, or digging up the garden (Sutton and Taylor, 2022). Despite this, sharing a home with non-human animals also provides opportunities. Animals can and do show us how they would like to use space (Philo and Wilbert, 2000; Smith, 2003; Sutton and Taylor, 2022), and by paying attention to their placement of items, arrangement of furniture, and movement through space, we can arrive at a less human-centric understanding of what matters to our animal companions, how they might see the world, and how we, in turn, might relinquish some of our control in order to be better companions to those with whom we share our homes.

The COVID-19 pandemic has made anthroparchal relations and the social construction of animals as pet commodities more starkly visible than ever before. As contact with human others was restricted, more emphasis was placed on pets to meet the social, emotional, and physical needs of their human owners (e.g. Young et al. 2020). This was also an era marked by lockdowns and quarantine orders, which saw many people confined to their homes for extended periods of time, and thus experiencing the lived realities of navigating multispecies lives more acutely than they otherwise would. This, then, provides a fruitful opportunity to explore how human–companion relations have manifested during the pandemic and demonstrate the values of using a sociological lens to interpret patterns of multispecies relations through the key theories outlined above. This intertwining of sociological concepts and data uncovers the potential opportunities offered by these periods of home-dwelling for reimagining multispecies companion relationships. However, opportunities to rethink human–animal relationships only exist if they can be imagined by human social actors – necessitating an authorizing environment and media institutions that support this reflexive engagement. As will be shown in the analysis below, such an environment does not exist, with Australian media instead perpetuating anthroparchal norms that centre humans and minimize any critical reflection on power relations. But first, the following section examines the role and significance of the media in manufacturing these norms, in order to demonstrate why researching media depictions is relevant to understanding human–companion animal relationships generally and during the pandemic specifically.

Media: the anthroparchal propaganda machine

Critical perspectives on mass media highlight the role this institution plays in shaping societal imaginaries and fostering accompanying norms and values. Functions of the media include 'manufacturing consent' for political and economic systems in society (Herman & Chomsky 1988), creating moral panics (Hall et al. 2017) and fostering collective emotions through the framing of events (Altheide 2010). Mass media represents a useful tool for maintaining the anti-animal status quo, as it performs 'a system-supportive propaganda function, by reliance on market forces, internalized assumptions, and self censorship, and without overt coercion'(Herman & Chomsky 1988, p. 306). Núria Almiron's work (2015, p. 27) has highlighted how industries that depend on non-human animal abuse 'enjoy an exclusive position to influence the manufacturing of consent for this abuse' through lobbying, provision of research funds, and provision of expertize to media, public opinion, and governments. Matthew Cole and Kate Stewart's (2014) critical work highlighted the particular normalization of animal use/abuse of children that is perpetuated through a range of social influences, including media. Considering the use of non-human animals as pets is a 3.7 billion dollar industry in Australia alone (IBIS 2023), and 69% of Australian households already include at least one pet, there is a heavy social and economic investment in the continued commodification of animals for companionship.

Method

In order to explore the discourses around companion animals in Australia during the COVID-19 pandemic, an analysis of Australian print media was conducted. Articles were sourced through the Australia & New Zealand Newsstream database, using search terms that would include species most likely to be used as companions with broader terms such as 'pet' and 'companion animal' to include other species (see Table 2.1). Given the clear tendency for Australian households to favour cats and dogs as companion animals (Wilkins et al., 2020), these are the only species specifically named in the search terms. This may have limited the results obtained; however, the inclusion of catch-all phrases such as 'Pet(s)' and 'Companion Animal' is likely to minimize any limitation by capturing non-dog and cat pets regardless.

I chose to focus on articles published from December 2019 onwards, as this covered the period when COVID-19 first came to the attention of the Australian public until the time of analysis, allowing me to track patterns in the depiction of pets and COVID over time. My search terms returned 6633 articles, all of

Table 2.1 Terms used to source articles

Boolean terms	*Australia AND (Covid OR Coronavirus) AND (Pet OR Pets OR 'Companion Animal' OR Cat OR Dog OR kitten OR puppy)*
Date limit	1/12/2019–24/11/2020
Total articles downloaded	6633
Minus duplicates and irrelevant articles	– 6516
Articles remaining	117

which were reviewed to eliminate duplicates and irrelevant articles. Articles were judged to be irrelevant if they did not discuss both animals used as companions and COVID-19, and the Australian tendency to name sports teams after non-human species resulted in a significant number of sports-related articles responding to the search terms (all of which were eliminated). This left 117 articles in the sample which were thematically coded using the NVivo12 software.

The media review focused solely on news articles and opinion pieces, rather than correspondence-type pieces such as letters to the editor, which may be another fruitful site of examination to gain insights into some of the public responses to the patterns of discourse described in this article. While these neglected sources may pose some interest, I chose to limit the sample to focus solely on the depictions of pets during COVID-19 in curated, original articles to demonstrate what news creators felt were the most 'newsworthy' topics during the pandemic.

Articles were coded according to the focus of the article (topic discussed), whether they were human- or animal-centric and the species discussed. The aim here was to highlight patterns in who, what, and how articles on animals used as companions were depicted in the context of the pandemic. The results of this analysis are discussed in the next section.

Animals as pet commodities: What media depictions can tell us about pets during the pandemic

Species

Aside from humans who were mentioned 100 per cent of the time, the articles were overwhelmingly dog-focused (54 per cent), followed by cat-focused articles (18 per cent) and those that referred to 'pets' generally (18 per cent). As Table 2.2 shows, other species were mentioned; however, this was significantly less often.

Topic

The articles covered a range of topics (n = 25), which were then coded into five overarching themes: commodification; navigation of everyday life; pets' impact on humans' health (physical and mental); interspecies infection risks; and social issues faced by animals used as companions during the pandemic. As can be seen in Figure 2.1, the themes most written about were commodification (36%) and navigation of everyday life (35%). These themes are discussed further below.

Commodification

Articles coded within the theme of 'commodification' were focused on the acquisition or sale of animals to be used as companions (n = 3 6), puppy 'scams' (n = 17), services built around pet commodities (n = 6), relinquishments or 'returning' unwanted pets (n = 4), goods and accessories (n = 3), and

Table 2.2 Number of articles mentioning each type of pet

Species	Number of articles
Bird	3
Cat	29
Chicken	3
Crayfish	1
Dog	84
Ferret	1
Guinea pig	1
Horse	2
Human	117
Pets generally	29
Rabbit	2
Reptile	2

Commodification Infection threat
Navigation of everyday life Social problems
CA impact on Human physical/mental health

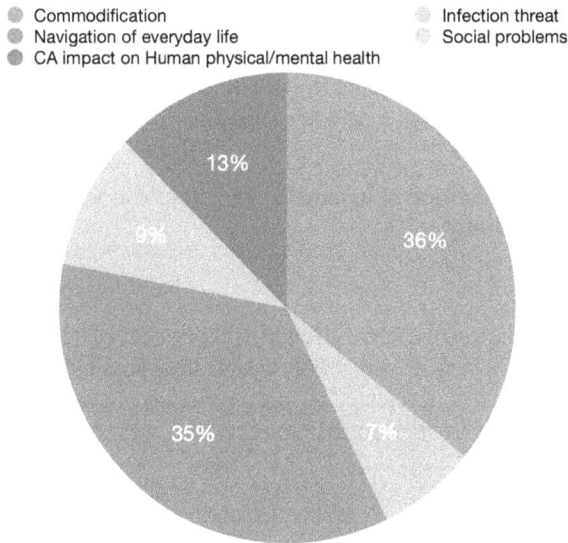

Figure 2.1 Most common themes found.

animal theft (n = 2). The vast majority of these articles were concerned with the increased demand for pets during the pandemic – for example:

> Diamond Valley Kennels owner Lisa Howlett says her operation, with 32 breeding dogs, is the state's biggest kennel supplying cavoodles and spoo-dles and she simply can't keep up with an insatiable demand for the dogs. At the peak of demand this year, there were 635 people on her waiting list for puppies.

(17 October 2020)

People are still going clucky for backyard chickens across the Central Highlands as lockdown restrictions continue to draw on.

('In a flap for pet chooks, egg care', 30 June 2020)

Lucky cats, dogs, birds and even bunnies across the state have been given a second chance at life as RSPCA NSW has recorded a 30 per cent increase in adoptions compared to this time last year.

('Animals thriving as lockdown families turn to pets', 12 April 2020)

The next highest subcategory in this theme were articles reporting on the rise of 'puppy scams' – the deceitful procurement of funds in return for puppies that did not in fact exist, a practice that seemingly boomed during the pandemic:

If a cute puppy offered online sounds too good to be true, it probably is. That is the message from the Australian Competition & Consumer Commission (ACCC), which estimates that Australians have lost almost $300,000 to puppy scams during the coronavirus pandemic.

(19 May 2020)

Navigation of everyday life

Articles coded under the 'navigation of everyday life' theme focused on recommendations for activities to be done with animal companions during COVID-19 (n = 25), accessing veterinary care (n = 17), welfare concerns for pets during the pandemic (n = 19), and the everyday impact of COVID-19 on pets (n = 8). A small number of articles considered arrangements to board (n = 1), foster (n = 2), or pet-sit (n = 2) animals used as companions during the pandemic:

A fun activity for the family was to dress up in costumes, so the puppy gets used to different people and looks. 'Practise your acting skills wearing wigs, hats, sunglasses, pretend to be a delivery person, use a cane or a walker, get bundled up in wet weather or winter gear,' Dr Zito said.

('Perfect puppies still need training', 12 April 2020)

Amelia Fidler was desperate to get her miniature dachshund Gus to a specialist vet clinic for urgent treatment. The 12-year-old dog was booked for radiation for an oral melanoma that was growing rapidly – an operation not on offer in Tasmania. However, because of existing COVID rules she could not cross into the state.

('"Save my dog": Border emergency for mini sausage',
17 November 2020)

Impact of pets on human health

Articles touting the positive impact of animal companions on humans' mental and physical health made up a small, but not insignificant, portion of the sample (13 per cent). These articles highlighted the increased physical activity, comfort, and amelioration of (human) loneliness that the use of other animals facilitated for the humans who owned them during the pandemic:

> Pets are a great source of comfort and companionship, can reduce anxiety, stress, and depression and encourage playfulness and exercise.
> ('Animal adoptions surge in wake of bushfire recovery, COVID lockdowns', 14 August 2020)

> With everyone cooped-up inside, it kind of made sense to allow it to happen now.' His son Riley said the energetic kitten 'takes my mind off things that are going on these days.
> ('Coronavirus restrictions see demand for pets surge as shelters issue warning to prospective owners', 4 April 2020)

Social issues

A small number of articles considered social issues faced by animals used as companions during COVID-19. The most significantly cited of these (n = 13) were travel issues, with border restrictions impacting animals' ability to access healthcare, and/or being stranded away from their families when they were either left behind in other locations before borders closed or sent ahead without their families if the humans were delayed due to border restrictions. Two articles described the increasing public nuisance of pets' presence in neighbourhoods as their owners spent more time out in public with them and one article highlighted the increased risk of companion animals facing family violence as domestic and family violence rates increased significantly for humans during periods of lockdown:

> Tough travel restrictions in Western Australia's far north have not only separated people, but also left some pets stranded thousands of kilometres from their owners.
> ('COVID-19 travel restrictions leave horses stuck thousands of kilometres from owners', 24 April 2020)

> When coronavirus struck, Shannon's employer recommended he return home immediately but while preparing to leave, Shannon learnt all pet imports from Malaysia have been suspended.
> ('Expats trying to bring pets home face difficulties', 18 May 2020)

Interspecies infection risk

Finally, 11 articles focused on the risk of interspecies infection with COVID-19. These were more evident earlier in the date period, with the majority of the articles appearing in March and April 2020. As the pandemic wore on, concerns about interspecies infection transmission began to focus on other zoonotic diseases, such as *Salmonella*, obtained through increased avian contact following the increase in purchases of chickens to be used as companions:

> if infection has occurred it has been human to animal and not the other way around. A trend that does seem to be emerging through laboratory testing is that cats and to a lesser degree, ferrets, appear susceptible to the illness.
> ('Are my pets at risk of catching COVID-19?', 18 April 2020)

> 'To date, there is no evidence that companion animals play a role in the spread of this human disease or that they become sick if they are exposed to the virus,' said an RSPCA spokesman.
> ('"No risk of pets spreading COVID-19," says RSPCA',
> 25 March 2020)

Focus

Most of the articles written about pets and COVID-19 in Australia were human-centric (72 per cent). These articles reflected on the experiences around commodifying, cohabitating, and navigating interspecies life *for humans*. For instance, some articles describing the social issue of animals being separated from their human families due to border restrictions focused on how this phenomenon affected the humans involved:

> Kerri-Ann Hudson moved to Kununurra from Perth last year; it was too late in the season for a transport company to bring her horse, Gunna, so she boarded him 3,000 kilometres away at a property outside Perth. She started planning months ago to bring him to Kununurra once trucking companies began operating again, but when it came for Gunna to be loaded onto the truck, the company told her it wasn't possible to get to Kununurra. 'I was pretty bloody devastated,' she said.
> ('Turned away at the border: COVID-19 travel restrictions leave horses stuck thousands of kilometres from owners', 24 April 2020)

> When Dean White, 46, moved to Britain three years ago for work, he did not imagine returning to Australia without his family dog, Luma. But as borders are shut to stop the spread of coronavirus, the prospect of having to leave his pet overseas is becoming all too real. 'I am heartbroken,'

Mr White said. 'It is like someone being told their child is locked out and can't come home despite Scott Morrison saying let's bring every Australian home.'

('Family heartbroken as pet dog stranded in
UK', 24 March 2020)

Human-centric articles did not consider the experiences or perspectives of animals used as companions – for instance, in the travel articles cited above, there was no mention of what the experience of familial separation and containment in an unfamiliar environment for an undetermined period of time might be like for the horses or dogs involved.

Some articles did consider other animals' experiences *to an extent*, describing these alongside depictions of human experiences. These articles were coded as having an ambiguous focus and accounted for 19 per cent of the articles in the sample:

> Working full-time from home for a major bank since March, the dogs have been enjoying longer morning walks, frequent outdoor breaks, three daily meals and, of course, 'cuddles on the couch for most of the day'. They've loved it, Pisani-Forde says, and so has she. But she's expecting to return to the office in June and she's feeling a little worried, saying she fears they have become 'velcro dogs'.
>
> ('Prepare your pets for your return to the office', 31 May 2020)

Finally, a small proportion of the articles in the sample were animal-centric (9 per cent). These highlighted the increased injury risks (from walking and being trapped in elevators), imposition and potential infection risk animals faced as a result of the pandemic, and increasing amounts of time spent with humans in their spaces. Importantly, these articles did not bring the discussion back to humans – for instance, articles discussing interspecies infection risks in an animal-centric manner were focused entirely on the need for animals to be protected from humans who were infected with COVID-19 to prevent the transmission of illness, without ever discussing human experiences:

> Australia's 'fur babies' have been visiting the vet twice as much during COVID-19 with a spike in conditions linked to overexercising, including torn cruciate ligaments … Dr Kovac said young puppies, older dogs and pooches with existing medical conditions were 'most at risk' of overexertion injuries.
>
> ('Corona's no walk in the park for pups', 7 August 2020)

With so many of us now encouraged to work & study at home, our pets are being forced to live with a new normal – one in which they are not left alone for eight or more hours a day. For the extroverted & jumpy dog, this may be like a golden age. For the reserved & sleepy cat, it might feel closer to a nightmare.

('Has COVID-19 seen your pet acting like a complete weirdo? An animal behaviour expert explains why', 1 May 2020)

COVID-19: Reflections on companion animal narratives

Although the small proportion of articles either partially or wholly centring on non-human animals' wants and needs is at least a step towards confronting human-centrism, overwhelmingly, the Australian print media reviewed reflects anthroparchal norms in centring human wants, needs, and perspectives in their reporting of companion animals. Coupled with the significant focus on the commodification of non-human animals as pets, the narratives studied largely reinforce discourses around non-human animals as a buyable source of companionship, an object for sale that, as Rosemary-Claire Collard (2014) states, is lively but never fully alive. This is further supported by the articles promoting non-human animals as beneficial for *human* mental and physical health, with the language here reflecting expectations that pets will be entertaining, stress relieving, and almost able to be prescribed to soothe the ails of living through a global pandemic. The message in these themes is that pet commodities exist for the enjoyment and benefit of human owners, requiring little in return outside of the initial cost of procurement and waiting period as non-human infants 'mature' enough to be separated from their families, given the high demand reported.

Also evident is the tension between these commodification discourses and the navigation of human–companion animal relationships in everyday life. Here, stories centred on activities to be done with pets and the impact of COVID-19 on non-human animals. Although often human-centric, human love and affection for pets is evident in these stories, the combined effect of the commodification and everyday life themes clearly illustrate Fudge's (2008) dual contradictory narratives. But one cannot escape the fact that the love described in these sources *was* predominantly human-centric – where companion animals faced separation from their human families, inability to access healthcare, and other hardships, these were represented as negative events *for human owners*. Any such consideration for the lived experiences of non-human animals was largely, and loudly, missing.

It is noteworthy that where stories did focus on non-human animals, these were largely reporting the negative impacts of human actions whether it be over-exercising, regulations that impeded access to healthcare, or the disruption of non-human animals' everyday lives that were suddenly, constantly infiltrated by humans. These again suggest a failure of owners to consider non-human animals' wants, needs, and experiences of their relationships with humans during the pandemic.

Conclusion: Sociological lens for companionship beyond consumerism or opportunities to rethink companionship?

The findings of this project indicate that humans may have missed an opportunity during the COVID-19 lockdowns to critically examine their relationships with non-human others. The doubling down on rampant consumerism evident in the significantly increased demand for pets, waitlists, and the rise of 'puppy scams' indicate that the use of non-human animals as lively pet commodities is more desirable than ever, with little sign of this behaviour being challenged in the 'new normal'. Given this, a multispecies sociology is needed now more than ever.

As discussed earlier in the chapter, sociology as a discipline is well placed to situate these entanglements within broader anthroparchal power relations and critically reflect on the social construction of non-human animals for human use (Cole and Stewart, 2014; Cudworth, 2011). As we set about discovering what the 'new normal' looks like amidst pandemics, climate change, and natural disasters, the sociological toolkit has the potential to offer pathways to living 'less badly' with other beings in multiple ways. First, in highlighting the inescapably human-centric, asymmetrical power relations underpinning human–companion animal relationships, multispecies sociology is able to build a case for the need to reconsider human–animal companionship that relies on commodification. Second, the specificity of analysis enabled by utilizing the theories of anthroparchy and socio-discursive construction of non-human animals allows scholars to argue not only that these entanglements are exploitative and need to change but also to identify manifestations of this domination in everyday life, where it might otherwise go unnoticed. While the goal for critical scholars pursuing vegan, emancipatory sociologies *for* other animals must be total liberation, I argue that there is also merit in working to live less badly with lively commodities while we pursue these more radical futures (Cudworth and Hobden, 2017; Taylor and Sutton, 2018).

So what should we do with this knowledge? As discussed earlier in this chapter, critically considering how we navigate our homes with other animals, consciously facilitating animal agency and use of space might be a good place to start. This certainly applies to our companion animals; however, I argue that this is a project that needs to extend much further. If we are to truly reconsider multispecies companionship in the light of critical sociological perspectives on pet commodities, a large component of this will be a honest consideration of whether humans are good companions to the other beings with whom we share our living spaces. This means paying attention to (and certainly not harming) the animals we are encouraged to think of as non-sensible (such as spiders, rats, and flies), and being aware that our perspectives on those who are sensible are likely shaped by human-centric narratives designed to legitimate our human use of those beings. Educating ourselves about the wants and needs of non-human others so we might better create space to accommodate them is crucial to decentring the human in home spaces in ways that might align with non-human animals' requirements. Certainly, the anthropogenic drivers of climate change and (likely) future pandemics indicate a dire need to challenge the human-centric, anthroparchal thirst for exploitation, suggesting that this move to decentre the human cannot come soon enough. It may not be a total revolution (yet), but an attitudinal shift away from the human as the ruler of the home space and towards an understanding of humans as companions to other animals might encourage some humans to begin to notice, care, and cater for other animals and move one step closer to less-bad multispecies relationships that benefit us all.

Notes

1 Here I draw on Cudworth and Hobden's (2017) discussion of pursuit of post-human emancipation, in which they reason that choosing which multispecies entanglements to keep and which to reject is a key aspect of working towards multispecies flourishing. Their approach does not rely solely upon macro, overarching changes; instead, they advocate for a consideration of relational and contextual ways to live 'less badly' with plants and animals (Cudworth and Hobden, 2017, p. 150). Pet commodities are necessarily exploited to meet the needs and wants of humans, and within this context, I argue that it is more appropriate to discuss living 'less badly' than living 'well' which might imply such exploitation is not insurmountable.
2 For an overview of scholarly critique of human–pet relationships in the field of animal studies more broadly, see Sutton (2020).

References

Almiron, N. (2015). The political economy behind the oppression of other animals: Interest and influence. *Critical Animal and Media Studies*, 26–41. London: Routledge.

Altheide, D. (2010). Risk communication and the discourse of fear. *Catalan Journal of Communication & Cultural Studies*, 2(2), 145–58.

Anderson, B. (1991 [1983]). *Imagined Communities: Reflections on the Origin and Spread of Nationalism*. London: Verso.

Animal Medicines Australia. (2021). *Pets and the Pandemic: A Social Research Snapshot of Pets and People in the COVID-19 era*.

Archer, M.S. (1995). *Realist Social Theory: The Morphogenetic Approach*. Cambridge: Cambridge University Press.

Charles, N. (2014). 'Animals just love you as you are': Experiencing kinship across the species barrier. *Sociology*, 48, 715–30.

Cole, M. and Stewart, K. (2014). *Our Children and Other Animals: The Cultural Construction of Human–Animal Relations in Childhood*. Aldershot: Ashgate.

Collard, R.-C. (2014). Putting animals back together, taking commodities apart. *Annals of the Association of American Geographers*, 104(1), 151–65.

Cudworth, E. (2005). *Developing Ecofeminist Theory: The Complexity of Difference*. Basingstoke: Palgrave Macmillan.

Cudworth, E. (2011). *Social Lives with Other Animals: Tales of Sex, Death and Love*. Basingstoke: Palgrave Macmillan.

Cudworth, E. and Hobden, S. (2017). *The Emancipatory Project of Posthumanism*. London: Routledge.

Douglas, M. (1991). The idea of a home: A kind of space. *Social Research*, 58(1), 287–307.

Fudge, E. (2008). *Pets*. London: Routledge.

Giddens, A. (1984). *The Constitution of Society: Outline of the Theory of Structuration.* Cambridge: Polity Press.

Greenebaum, J. (2004). It's a dog's life: Elevating status from pet to 'fur baby' at yappy hour. *Society & Animals*, 12(2), 117–35.

Hall, S., Critcher, C., Jefferson, T., Clarke, J., and Roberts, B. (2017). *Policing the Crisis: Mugging, the State and Law and Order.* Bloomsbury Publishing.

Herman, E.S., and Chomsky, N. (1988). *Manufacturing Consent: The Political Economy of the Mass Media.* Pantheon Books.

IBIS. (August 2023). 'Pets and Pet Supplies Retailers in Australia' Industry report. Accessed at https://my.ibisworld.com/au/en/industry-specialized/od5128/at-a-glance on 30/08/2023.

Irvine, L. and Cilia, L. (2017), More-than-human families: pets, people, and practices in multispecies households. *Sociology Compass*, 11(2), 1–13.

Lefevbre, H. (1991 [1974]). *The Production of Space* (trans. Donald Nicholson-Smith). Oxford: Blackwell.

Philo, C. and Wilbert, C. (eds) (2000). *Animal Spaces, Beastly Places: New Geographies of Human–Animal Relations.* London: Routledge.

Smith, J.A. (2003). Beyond dominance and affection: Living with rabbits in post-humanist households. *Society and Animals*, 11(20), 181–97.

Stewart, K., and Cole, M. (2009). The conceptual separation of food and animals in childhood. *Food, Culture & Society*, 12(4), 457–76.

Sutton, Z. (2020). Researching towards a critically posthumanist future: On the political 'doing' of critical research for companion animal liberation. *International Journal of Sociology and Social Policy*. https://doi.org/10.1108/IJSSP-01-2020-0015

Sutton, Z. and Taylor, N. (2022). Between force and freedom: Place, space and animals-as-pet-commodities. In *Vegan Geographies*. New York: Lantern Press.

Taylor, N. and Sutton, Z. (2018). For an emancipatory animal sociology. *Journal of Sociology*, 54(4), 467–87.

Wilkins, R., Botha, F., Vera-Toscano, E., and Wooden, M. (2020). *The Household, Income and Labour Dynamics in Australia Survey: Selected Findings from Waves 1 to 18: The 15th Annual Statistical Report of the HILDA Survey.* Melbourne: University of Melbourne.

Young, J., Pritchard, R., Nottle, C., and Banwell, H. (2020). Pets, touch, and COVID-19: Health benefits from non-human touch through times of stress. *Journal of Behavioural Economics for Policy*, 4, 25–33.

3 Power, politics and representation in research with (other) animals in the 'new normal'

Nik Taylor and Heather Fraser

Figure 3.1 Image 1 from Vegan Wellbeing Project 2019–2020.

Source: Vegan Wellbeing Project 2019–20 (Fraser, Taylor, Stekelenburg & King).

Across the political spectrum, sociologists have always been interested in how societies form and operate and how power relations, social structures, status, social problems, collectivism and individualism all play a part. For many sociologists – especially those who identify as feminist – societal structures, processes and conventions are examined through the intersecting lens of gender, race/ethnicity, age, ability, sexuality and geographical location. However, sufficient sociological attention has not been given to non-human animals and the natural environment, despite ongoing calls from feminist sociology to centre climate change (Fox and Alldred, 2020; Kaijser and Kronsell, 2014; MacGregor, 2009, 2014) and analyses of species and animals (Taylor, 2011). Now is the time to do so, given the implications of the climate emergency and the COVID-19 pandemic.

The climate change emergency and COVID-19 have made it abundantly clear to many of us that our current relationships with other animals are not only

DOI: 10.4324/9781003257912-5

unsustainable but also dangerous. The killing, sale and consumption of other animals and their body parts/secretions generates risks for humans, as well as creating misery and pain for countless animals annually (Brozek and Falkenberg, 2021). It is well established that big agribusiness fuels are expanding meat and dairy markets, which creates multiple and irreparable harms to the landscape, water and air quality (Schally, 2017) and that, in addition to COVID-19, there are multiple instances of diseases jumping from animals to humans due to our farming and consumption practices. For example, the coronavirus Severe Acute Respiratory Syndrome (SARS), which infected an estimated 5,000 people, was traced to several humans working in live animal markets in China, with there being 'little doubt' (Reddy and Saier, 2020) that the virus jumped from bats to other animals and then on to the humans who interacted with those secondary animals in the markets.

However, it is striking that among the countless space given over to commentary on climate change and COVID-19 (e.g. Davis, 2021), relatively little of it focuses on meat consumption or our relations with other animals more generally. Instead, we see technical solutions proposed that provide a smokescreen for the systemic nature of both climate change and pandemic reproduction. So resistant are we to the change required to address the climate emergency, and so reliant on technical solutions, that we see countries such as Australia banking on technologies that have yet to be developed to fix problems (Wu 2021). Similarly, rather than challenge the notions of human supremacy and carnism (or the ideology of meat eating (Taylor and Fraser, 2017), mainstream injunctions are to improve handling and containment of 'risky' animal species and contact zones between them and humans (Kochevar and Werneck, 2021). What is missing is commentary on how the destruction of 'nature' contributes to such outbreaks, plummeting biodiversity and escalating climate crises (Carrington, 2020). This very absence of critical commentary helps to normalize the legitimacy of big agribusiness and its carnist markets. Even at this critical juncture, and with all the evidence of their harms, talk of veganism is likely to be considered 'radical' or even 'unhinged' (see Taylor et al., 2022).

How we, as scholar advocates and/or activists, challenge this normative understanding of human superiority becomes crucially important and is considered by many working in critical animal studies (Taylor, Struthers Montford and Kasprzycka, 2021), feminist animal studies (Cudworth, 2016) and the humanities and social sciences more generally (e.g. Peggs, 2013). It is also a question that sociologists should be considering, given that our discipline is – at least nominally – concerned with the ways inequality manifests in social life, institutions and knowledge. However, as we argue below, sociologists have all too often refused to consider other species due to the lingering, and we argue problematic, humanism that pervades our discipline (and so many others). Nevertheless, we believe sociology can contribute to the unravelling of the human superiority claims that have led to the current state of crisis. But to do so, it/we must confront our epistemic speciesism, and this includes acknowledging the anthropocentrism that is built into our theories and our methods.

As we discuss below, post-humanism and its sociological antecedents of actor–network theory (ANT) and science and technology studies (STS) offer a starting point here. The aim of this chapter, then, is to explicitly consider how the politics of method and representation contribute to the acceptance of the oppression and exploitation of other animals. We then turn to consider how this can be disrupted by arguing that the new normal necessitates an alternate approach that is species-inclusive. Underpinning this is our intersectional understanding of multispecies relations based on the recognition and respect for difference, reciprocity and mutual care. Our impetus is to advance the rights of all oppressed groups – including non-human animals – in a time described as 'the new normal', but that we see as anything but normal or healthy.

On a daily basis, the number of animals abused and tortured globally to meet human 'needs' is in the billions. According to one source (www.adaptt. org/killcounter.html), in the time it has taken to write these sentences 82,049 marine animals, 41,840 chickens, 2062 ducks, 1134 pigs, 781 rabbits, 630 turkeys, 486 geese, 470 sheep, 315 goats, 266 cows and calves, 59 rodents, 57 pigeons and other birds, 21 buffaloes, 15 dogs, four cats, four horses, three donkeys and mules and two camels and other camelids have been killed. As the source makes clear, this is an underestimate:

> These are the numbers of animals killed worldwide by the meat, egg and dairy industries since you opened this webpage. These numbers do NOT include the many millions of animals killed each year in vivisection laboratories. They do NOT include the millions of dogs and cats killed in animal shelters every year. They do NOT include the animals who died while held captive in the animal-slavery enterprises of circuses, rodeos, zoos, and marine parks. They do NOT include the animals killed while pressed into such blood sports as bullfighting, cockfighting, dogfighting, and bear-baiting, nor do they include horses and greyhounds who were exterminated after they were no longer deemed suitable for racing.

How are we to make sense of these staggering numbers? Should we try to make sense of them? Or are such attempts likely to be a way of excusing them? Given such large-scale slaughter of animals occurs largely unconsidered and unchallenged, what are we to make of the impotence that such data carry in changing the outcomes for animals? What role do data (quantitative and qualitative) play in the production of knowledge about animals and our abuse of them? Can academic work contribute to the goal of making individual animal lives better? Should it? What are the consequences of adopting an overtly activist and politicized stance toward our work? What is, or can be, the role of sociology in activist scholarship for other animals? These are just some of the questions that guide this chapter, and our work more broadly, as we consider the ramifications of not just studying other animals but also the need to advocate for/with them in the crisis-ridden new normal.

How research is framed, measured, qualified and conducted has shaped and continues to help shape, rationalize and justify human treatment of (other) animals. In this chapter, we ask whether and how the study of non-human animals affects our comprehension of social science epistemologies. Conversely, we also ask whether and how sociological epistemologies might affect the way we see – and ultimately treat – other animals. Aware that including non-humans in our scholarship has the radical potential to subvert traditional sociological paradigms that rest upon neat divisions between 'the' social and 'the' natural (Latour, 2007), we ask what the politics of turning our attention to the study of other animals might be. We begin by outlining our starting points and underlying premises. We then consider the post-human turn in the academy and its consequences for studying human–animal relations sociologically before discussing the politics of representing other species and what this means for them while considering the epistemological implications bound up in attempts to make other species 'known'.

Our starting points

Throughout the current pandemic, we have had many conversations (primarily with each other, but with other academics as well) about the purpose of academia and our role within it. As academics who work at the margins due to numerous factors (vegan, feminist, working class and animal advocates), this question was never far from our thoughts, conversation and writing, but the pandemic gave it added urgency. We talked about why we remain in a system that historically has supported huge inequalities, including playing its part in the normalization of animal abuses across the board. Furthermore, rather than being a chance to rethink and revise how we do things in academia, the pandemic seems to have encouraged many in the system to double down on the neoliberal managerialism that aims to turn it into a knowledge-for-profit business. Complicity with at least some of this cannot be avoided when working in this system; so, we talked about whether we should stay and, if so, how we might challenge the status quo, and particularly use our knowledge and status to advocate for other animals. In turn, this led to (yet more) conversations about the nature and limitations of scholar advocacy. Ultimately, this led us to a consideration of our position *vis-à-vis* sociology and its power to create change. We openly considered our rocky relationship with sociology, which has ranged from a strong belief in its abilities to challenge power structures and be one part of the push for change, to a more cynical disregard for its current academic face that often feels careerist, conservative, niche and part of the problem. This is hardly surprising given that 'the discipline came into existence as a "provider of facts" to help political rulers rule'. It laid claim to scientific status on a par with the natural sciences, for an indivisible 'science' that played no small part in the fact that 'Its dominating motifs are the separations between knowers and what/who is known, subjectivity and objectivity, science and nature' (Stanley, 1990, pp. 10–11).

In many ways, then, while we do believe sociology can – and should – be a driver in challenging the anti-animal status quo, it was/is conceived in such a way that this is an extremely difficult – if not impossible – ask of a discipline that thus far has rejected serious attempts to overhaul its dominant epistemological orientation to make room for 'others': here we are thinking of, for example, feminist challenges and those from environmental sociology (e.g. Canan, 1996; McCann, 2016; Pellow and Brehm, 2013; Stacey and Thorne, 1985).

Despite this, we are still here, still working in academia and our continued presence goes beyond the need for a pay packet. Our last conversation while writing this chapter turned to why this is the case, despite the misgivings we have about academia generally and sociology specifically and their capacity to create change. The short answer is that we believe in the *potential* transformative power of knowledge. This isn't a naïve or straightforward belief; we are well aware of the complexities involved in advocating for and creating change. But we do believe that sociological knowledge at least has the capacity to promote change – in the classroom and outside it. In part, this is because of one of the discipline's key injunctions that we use the sociological imagination to see the links between personal experiences and wider societal structures (Mills, 1959). In part, too, it is due to the rich history within feminist sociology of addressing the role of epistemological power in society (Dallimore, 2000; Harding, 1987; Houston and Kramarae, 1991; Stacey and Thorne, 1985; Stanley, 1990; Stanley and Wise, 1993). Taken together, these lead us to the conclusion that sociological knowledge can be used to promote change and, in turn, that this demands that academics must also be advocates. There is no room for claims about neutrality in a world that is slowly dying – if there ever has been. And there is certainly no room for complicity in the face of structures and institutions (such as academia) that often support the damages being wrought on our environment and the beings that live in it. Instead, there is a need for action, a need for praxis. Defined as 'a feminist commitment to a political position in which knowledge is not simply defined as "knowledge *what*" but also as "knowledge *for*". Succinctly the point is to change the world, not only study it' (Stanley, 1990, p. 14). Adopting such a feminist sociological position opens up the transformational power that academia has *in potentia*. But that power faces barriers, and for other animals, these are multiple – especially, as stated above, from within a discipline that is largely anthropocentric and humanist. This means that not only are other animals left unconsidered in our theories and concepts – or, perhaps worse, only considered in terms of their importance to humans – but they are excluded from our methods in ways that normalize our lack of concern with/for them. This we name 'epistemic speciesism' and aligned 'methodological speciesism'. And this is the focus of the remainder of this chapter.

We work from the premise that all research methods are political (e.g. Law, 2004) insofar as they all frame research questions in particular ways, are deployed for preordained purposes and produce particular kinds of findings disseminated in socio-political contexts. This applies across the divides of the physical ('hard') or social ('soft') sciences, and irrespective of whether (or not) the research is tied to commercial interests. Wittingly or otherwise, researchers across disciplines and employment arrangements interact with (subscribe to, support, criticize and/or oppose) dominant world views about problems, their causes and possible remedies:

> The dominant worldview is not just one way to view the world; it is posi-
> tioned as the most legitimate way to view the world and as such, it is
> difficult to resist … Under the dominant paradigm of positivism, quan-
> titative measures of rigor and validity are the 'gold standard' through
> which 'proof' is established and information attains the status of
> knowledge.
>
> (Strega, 2015, p. 121)

We do not take an antagonistic approach to quantitative research nor do we privilege it over qualitative research. However, we do not support the idea that quantitative research is the 'gold standard' of 'proof'. For us, validity is a contestable concept (Dallimore, 2000). We ask ourselves whether the evidence is accurate and reveals important 'truths' and whether the attendant claims made are valid. Conventional (scientific) definitions of validity usually refer to replicable testing (often in laboratories), where variables can be controlled. For those of us who study humans and other animals who live in complex, multidimensional contexts involving variables that produce interactions that cannot be controlled or usefully disaggregated, we know that this kind of validity is limited, if not irrelevant. Our understanding of validity relates more to Dallimore's (2000) concept of trustworthiness. This involves indicating the research design and process, including any subsequent changes. It also rests upon ensuring that ordinary people's testimonies are not pushed aside in preference for those made by experts and professionals and that efforts are made to engage 'inconvenient samples' or 'hidden populations' in the production of knowledge. As we suggest below, these 'hidden' populations and samples of 'inconvenience' may relate to the study of animals, particularly the everyday relationships shared between humans and companion animals.

Most importantly, we do not accept that it is possible to achieve complete neutrality in research. Complete neutrality and objectivity are a chimera because *people* undertake research; individuals and groups that have their personal, sociocultural, professional and disciplinary ways of knowledge production. For instance, we are two middle-aged, White, Western, feminist academics from working-class backgrounds who have expertise in qualitative research and who care about the treatment of animals, not just humans. Our approach to research, the language we use, the questions we pose and the conclusions we draw will be different from those who purport no such interest in the fair treatment of animals or who oppose any change to humans dominating (other) animals. Either way, the production of knowledge is always a political endeavour shaped by wider forces of personality, networks, institutions and cultures (e.g. Drew and Taylor, 2014; Fraser and Taylor, 2016; Taylor and Fraser, 2019).

If neutrality is not a possible starting point, then the goal of 'good research' is to produce knowledge that seeks to benefit society, and we argue that – particularly in current 'new normal' times – this simply must include other animals and the environment. In part, this is for the benefit of humanity, given the clear deleterious effects the appalling (dominant, Western) treatment of other animals and the environment to date has had on our species and the world we share with others. But it is also with the aim of 'benefiting' other animals by making explicit the systems and structures that work to oppress them, with the ultimate aim of ending that oppression. Such research is conducted in organized and thoughtful ways that acknowledge the perspective-taking of researchers, and it gives due thought to the contexts (time, space and place) within which the study is being undertaken. 'Good research' is fair insofar as it does not conceal subjectivity or hide confusion, contradiction and mess (e.g. Fraser and Taylor, 2016; Hamilton and Taylor, 2017; Strega, 2015). It is research done by those who are willing to embrace their partiality and their subjectivity. As feminist scholars have made clear, this frank approach is itself political because it involves a refusal to hide behind an idealized version of the research process that rests upon a rationalized, masculinist and liberal humanism (e.g. Harding, 1987; Smith, 1987; Stanley and Wise, 1993). By articulating it in such a way, anyone who follows this line of thinking needs to make clear their normative, political and ideological commitments in order to mediate against the idea that research is 'neutral' when claims of neutrality do little more than support the (in this case anti-animal) status quo (Drew and Taylor, 2014). Where possible, such an approach lays bare the underlying premises and ideological, theoretical and epistemological underpinnings, allowing readers to adjudicate for themselves the likely accuracy and relevance of findings.

These abstracted questions about validity, trustworthiness and the production of knowledge may seem a long way away from the needs and rights of animals, but they matter a great deal. As we have shown above, even 'hard' data with 'big numbers' can have limited potential to persuade people to collectively change their approaches to animals. Consider the relatively available information on state-sanctioned practices in the meat and dairy industries and in the tourism and entertainment industries such as racing, hunting and fishing (e.g. Orzechowski, 2022). When groups are so socially devalued, compelling evidence of large-scale cruelty and enslavement can be imagined away – and for good reason. Selective amnesia, sometimes culturally induced, is a coping strategy used by many, if not all, to deal with distress and contradictions. Animal lovers who eat meat but refuse to hear of the processes used to produce their steak, smokers who understand the health risks but continue to smoke and corporations that knowingly destroy habitat are just three examples of how people can set aside knowledge they do not dispute as worthy or true to pursue other priorities. This means it is not only producing valid, representative and trustworthy knowledge that is needed; we also require clarity – or at least an idea – about what to do with this knowledge post-production.

Intentionally or otherwise, the marginalization, oppression and violation of non-human animals is extensive. While the societal and ideological processes that marginalize and silence other species are complex (e.g. Adams, 1990; Cudworth, 2015; Kemmerer, 2011; Peggs, 2013; Sutton and Taylor, 2019), one key part is their exclusion from our intellectual (epistemological and method- ological) sphere of concern. Here, then, we consider what kinds of methods we might use to understand the full range of relationship possibilities (political, emotional, material and corporeal) that are evidenced in human–animal rela- tions. Finally, we consider what methods and language we still might need to invent to 'capture' animal interests, and whether domination, anthropomor- phism and paternalism are inevitable as we navigate our way through the new normal.

Animals, the academy and the post-human turn

The last few decades have seen the growth and consolidation of the field of animal studies (also variously called anthrozoology, human–animal studies and critical animal studies).[1] As a multidisciplinary field, it is the study of the rela- tionships between non-human and human animals. The field focuses on both the historical and contemporary ways in which animals feature in human lives, cultures and societies (DeMello, 2012; Taylor, 2013). This includes the study of fictional, real and symbolic relationships between the species. The field is both interdisciplinary and multidisciplinary and involves work from the natural sci- ences as well as the arts, humanities, social sciences and sociology (see Taylor and Sutton, 2018 for an overview).

Across disciplines and methods, animal studies researchers are informed by a general compassionate ethic for animals and are often critical of the ways that animals are treated in human societies, which they ordinarily hope to improve (DeMello, 2012; Taylor, 2013). One part of this involves investigating the ways animals have come to be socially and symbolically constructed (e.g. Arluke and Sanders, 1996; Cole and Stewart, 2014; Sutton and Taylor, 2019), prompting some to criticize human–animal studies for becoming too abstract and esoteric – for obscuring, if not abandoning, the plight of real animals. In response to this, the sub-field of critical animal studies has developed, which explicitly links scholarship to political activism on behalf of animals (Best, 2009; Taylor and Twine, 2014).

Research in human–animal studies – critical or otherwise – involves an inherent curiosity about human relations with other animals – for example, the relationships many humans have with companion animals or free-roaming ani- mals that humans observe through regular visits (such as magpies, petrels, seagulls, dolphins and whales) (e.g. DeMello, 2012). Early research in this area tended to focus on the benefits to humans of the human–animal bond – for example, the mental and physical benefits of companion animal, or assistance animal, ownership (e.g. Beck and Katcher, 1996). More recent work has sought to draw attention to the various ways in which human–animal connections enrich the lives of all (e.g. Harrison et al., 2020).

Yet, there are much less pleasurable dimensions to human–animal studies, including studying the mechanisms and consequences of human objectification and the use and abuse of (other) animals. Some of this work focuses on the ways in which structures and institutions in society depend upon and encourage animal labour and exploitation – for example, slaughterhouses/ meat eating or animal experimentation (e.g. Cudworth, 2011, 2015; Gillespie, 2018; Taylor and Fraser, 2017; Torres, 2007; Wadiwel in Chapter 7 of this volume). This may take the form of analysing the lived interconnected oppressions of animals and disenfranchised human groups (e.g. Kemmerer, 2011), but also includes analysis of the (current and historical) symbolic and ideological ways in which animals are positioned as object-'others' and the kinds of treatment this encourages and legitimates (e.g. Adams 1990). Other work focuses more on the ways in which animals may be unwilling victims of dysfunctional inter-human relationships, such as those harmed within the context of family and domestic violence (e.g. Ascione, 1998; Becker and French, 2004; Riggs et al., 2018). Despite the diversity in this field of study, there is one factor common to all this work: it stresses the complexity of relationships between humans and other animals. This varied body of work has successfully demonstrated that animals are, and always have been, important to individual humans, structures, organizations and society as a whole.

Common to much of the work done in human–animal studies is the concept of anthropocentrism. For scholars in this field, anthropocentrism refers to the ways in which human interests and needs are prioritized over those of other species, and the planet more generally (see Boddice, 2011 for detailed discussion). It is here that the field often benefits from recent advances in post-humanism (e.g. Wolfe, 2010). Post-humanism, which grew from ideas about distributed agency first conceived in sociological studies of STS and ANT (Hornborg, 2017), has been embraced by many working with/for other animals. In part, this is because of its overt refusal of post-Enlightenment logical positivism and its elucidation of and attention to the power/knowledge processes inherent to binary/hierarchical thinking that are the outcome and pinnacle of liberal, individualist humanism (Taylor, 2011). Taken together, these offer one way to justify the inclusion of other animals in sociology, although we do note that the constant need to reiterate these justifications is now both tired and tiring, and ironically part of the broader processes that normalize their exclusion in the first place.

Studying animals: The politics of representation

Post-humanism, and its intellectual predecessor, ANT, opened the door to social scientists wanting to study the more-than-human (Law and Hassard, 1999). ANT theorists point out that the so-called social realm is in fact made up of entanglements – or 'assemblages', to use Latour's (2007) term – of human, technological, animal, material and object. Furthermore, those working in this paradigm argue that none of these has epistemic or analytical superiority. For the purpose of analysis done within ANT, animals, objects, humans and technology all occupy the same space because they contribute to social order and outcomes in equal amounts. This 'radical egalitarianism' (Chagani, 2014) gives permission to social scientists to study the more-than or other-than-human worlds *as part of their own*. By problematizing assumptions about human superiority, both ANT and posthumanism legitimate investigations into previously dismissed, if not ridiculed, areas such as emotions, human–animal relations, suffering and postcolonial and feminist ways of knowing that might stand against traditional versions. Taken together, this legitimates our interest in and concern with other species.[2]

Since humans dictate what constitutes intelligence and knowledge, all we know of (other) animals occurs through human perspectives. This has severe and usually negative consequences for animals, whose intelligence and knowledge are often ignored and whose value is determined by their importance to humans (Cole and Stewart, 2014). There are many examples of this. For instance, we protect some animals that we deem worthy, such as companion animals, for the intelligence we ascribe to them, or selected 'wildlife' that people find aesthetically pleasing, such as koalas, kangaroos and dolphins. Other animals such as cows, goats and sheep are valued not for their learning capacities, emotions or even their lives, but for their meat, milk and skins. So, we call them – and any other animal we wish to farm, eat and skin – 'livestock'.

Then there are species that we classify as threats to human prosperity. They become 'pests' that need to be 'managed', culled or exterminated, such as bats, corellas and cockatoos. Sometimes the same species that humans aesthetically admire are reconstituted as pests if their numbers swell beyond those deemed acceptable to (some) humans, or if they inhabit land, or eat plant material that humans wish to keep for themselves. Kangaroos, corellas and cockatoos are good examples of this. The euphemistic term 'cull' disappears the selective slaughter of a species, sometimes under the auspices of 'humane' treatment (Sutton and Taylor, 2019). Even among social progressives, questions about the legitimacy and audacity of humans casting the first and final vote about whether a species is 'over-abundant' are often absent.

Post-humanism is not without its problems, however. In many cases, it seems – somewhat ironically – to be more concerned with humans than animals (Kopnina, 2017). Here we are thinking, for example, of the post-humanist aim to move beyond the human organism in order to redefine humanity's relationships with other entities in the world (Wolfe, 2010). Similarly, feminist new materialism – arguably an intellectual offshoot of posthumanism – seems mired in problematic attempts to see *all matter* (living or dead) as ontologically similar. As Donovan points out, this does not move us far from the modes of thought that gave birth to the binary thinking that legitimates animal oppression in the first place. Arguing that new materialists' failure to consider that an 'it' may be an 'I' means that:

> In the end, then, all that is not human, in the New Materialist view, is an object, even if energized and actant. Such a view not only fails to obviate anthropocentrism but also actively allows a continuation of human manipulation and exploitation because 'It's' are still de facto held to be of lesser ontological status than 'I's', despite contentions to the contrary. We thus remain not far removed from Cartesian dualism.
>
> (Donovan, 2018, p. 259)

Despite these issues, when used by human–animal scholars, post-humanist ideas and their methodological legacies can go some way towards rectifying this assumed superiority and systemic domination of humans over all other animals (e.g. Blattner, Donaldson and Wilcox, 2020). For example, Sutton (2020, p. 386) discusses how her use of a 'species-inclusive methodological assemblage' allowed her to both challenge human-centric narratives of pet 'ownership' in her interviews with companion animal guardians and uncover everyday aspects of power negotiations between humans and animals that demonstrated animals had clear preferences about their use of space within the home. The strength of post-humanist ideas and methodologies lies in their refusal to accept that there is no other alternative. The challenge is how to work with post-humanist possibilities when human history and contemporary cultures are so humancentric. What knowledge might we value? Is it even possible to access the perspectives of other animals? If so, what research methods might be used?

Even though post-human attempts to 'know' other species are gaining ground, particularly in sociology (e.g. Law and Mol, 2008), we cannot escape human interpretations. This leaves us in a difficult position of wanting to understand but having that understanding limited by prevailing boundaries. This is often a source of frustration, as we do not have the epistemic toolkits needed to think or write about other animals except in ways that often contribute to their silence, their absence or their construction in our own – human – terms (Fraser and Taylor, 2020).

While post-human theories are starting to make some headway with this complex problem, methodological innovation is often lacking. Discussions remain at the abstract, conceptual level or they rely upon tried and tested methods that *do* write out and silence the species-different individuals we are trying to represent. One established method that does hold promise is ethnography (Hamilton and Taylor, 2012, 2017), although recent critiques note that despite claims/aims to decentre humans, *multispecies* ethnography to date has not engaged well with the exploitation of other animals (Gillespie, 2019, 2021; Kopnina, 2017). Ogden, Hall and Tanita (2013, p. 6) define multispecies ethnography as a project that seeks to understand the world as materially real, partially knowable, multicultured and multinatured, magical, and emergent through contingent relations of multiple beings and entities. Accordingly, the non-human world of multispecies encounters has its own logic and rules of engagement that exist within the larger articulations of the human world, encompassing the flow of nutrients and matter, the liveliness of animals, plants, bacteria, and other beings.

We share Kopnina's (2017) view that multispecies ethnography is often undertaken as an academic, intellectual exercise, devoid of engagement with animal liberation and oppression, in part because ethnographers refuse to start from the assumption that animals matter in their own right as opposed to through their value to humans. And, as Kopnina (2017, pp. 350–51) reminds us, there is no compelling reason why this should be (or at least remain) the case:

> Rather than asking why AR's [animal rights] should be considered, it is more essential to ask: why shouldn't they? What morality, other than 'might makes right' can justify non-human subordination? It is this type of questioning that needs to be evoked if the multispecies anthropology is to move beyond mere theorizing and 'ontological choreography'. The well-meaning but essentially apolitical academic excursions into multispecies ethnography may need to address multispecies injustice, suffering, and unidirectional violence.

Despite these limitations and the colonialist roots of ethnography aimed at describing, categorizing, ordering and policing 'natives' (Skeggs, 1999), it remains a 'politics of recognition' (Taylor, 1992). Taken to its critical apotheosis, it can be used in such a way that 'various sites of cultural contestation and everyday cultural practice … [are] … interrogated to better understand societal forces of power, dominance and change' (Foley, 2002, pp. 471–72) wherein it offers

> hope to create a practical, value-laden science that generates the knowledge needed to foster a democratic society and a critical citizenry … such knowledge production is part of a long dialogic consciousness-raising process. Such knowledge should have an 'emancipatory intent' (Habermas, 1971, cited in original) … Such a value-laden, didactic, practical social science differs markedly from a traditional, positivist notion of science.
>
> (Foley, 2002, p. 472)

Used carefully, judiciously and with deliberate attempts to problematize, challenge and ultimately change the subordinate position of animals, multispecies ethnography can become a powerful tool in the fight for animal liberation. As Gillespie (2021) argues, when merged with politicized, feminist approaches to knowledge, multispecies ethnography (MSE) (and, in her argument, multispecies autoethnography) can render 'the personal as political':

> Multispecies autoethnography can politicize intimate relationships of care, harm, power, and exclusion, and as such, can highlight aspects of human-animal relations that may be obscured in research that does not attend fully to the researcher's situatedness in a web of multispecies relations.
>
> (Gillespie, 2021)

But, first, any ethnography has to engage with the politics and ethics of representation.

Speaking for, representing and writing into 'reality' the lives of others is, at its heart, an arrogant endeavour as it assumes the 'other' *can* indeed be known and can be represented. The very act of earmarking a group to be understood suggests difference, usually without any caveat that this is simply 'different-than-the-ethnographer'. As others (e.g. Foucault, 1991) have convincingly argued, this delineation of self to other is itself a manoeuvre of power/knowledge and all too often can be used – even unwittingly or implicitly – against the group being studied. As Skeggs (1999, p. 34) points out,

> knowing the other through visual codes was central to the security of the powerful. In this way they could control physical space and generate their own sense of ontological security out of producing distance from supposedly recognizable others. Ethnography was used as one of the main technologies of the Enlightenment to generate classifications and knowledge about 'others'.

However, Skeggs (1999, p. 33) also argues that this is the appeal of ethnography – that it forces the ethnographer to constantly and consistently reflect on their values, ethics and approach; 'it is the impossibility of not being able to avoid the accusations of exploitation, colonial reproduction and the quagmire of representation and ethics that makes for a reflexive and vigilant researcher'. Used by reflexive practitioners, mindful of the power/knowledge locus of methodological attempts to represent others, ethnography can become a tool of liberation as opposed to oppression:

> As Paul Willis (1977) argues the role of ethnography is to show the cultural viewpoint of the oppressed, their 'hidden' knowledges and resistances ... The desire was not to know the other as other, or to study the other as a way of knowing oneself, but to understand how previously marginalized groups existed in circuits of meaning, cultural formation and structural location. It was to re-value the classification systems that had produced ossified positioning ... it was also a desire to question how particular theoretical orthodoxies had become established.
>
> (Skeggs, 1999, p. 36)

Conducting ethnography in order to study human–animal relations, however, presents an additional set of challenges – namely, how can ethnographers include the animals they purport to study (Gillespie, 2021; Hamilton and Taylor, 2017)? Unlike traditional methods such as surveys, ethnography does not automatically exclude animal participants, yet it is still the outcome of human actions and interpretations. Simply conducting ethnography in places where humans and other animals meet does not necessarily mean that animals are included, and it may be that, rather than needing a decentred human approach, we need an explicitly anti-anthropocentric one (Gillespie, 2021). An ethnographer aiming to consider this first has to tackle the issue that in the representational world of written reports animals *are* invisible and silent. While there may be ways to counter this, it will necessarily involve trekking into unknown terrain, which carries its own risks of being marginalized and ridiculed because the research does not abide by conventional standards (Becker, 1974). As Wilkinson (2005, p. 47) points out:

> In order to have their work resonate with a language of 'expert authority', social scientists all too easily arrive at the position of silencing the genuine voice of people who experience extremes of violence, material hardship and social upheaval. As a result, they ally themselves to the interests of those whose positions of power and privilege are maintained at the cost of the suffering populations.

While he is talking about the work of those attempting to understand human suffering, this is easily transferred to those attempting to understand the politics of animal suffering. In fact, it is possible that work focused on animals is even more negatively construed, as animals are considered even more unworthy of study than oppressed and marginalized human groups (Fraiman, 2012).

So those of us seeking to better understand human–animal relations, especially if we have an activist agenda that involves us using this knowledge to change appalling conditions for other animals, are often working in a double, or even triple, bind. Our own academic silencing and marginalization, and the societal silencing of animals, can continue through an uncritical acceptance of logical positivism into our methodological choices and constraints (e.g. Houston and Kramarae, 1991, for discussion on the multiple ways of academic silencing, in their case of women). This 'silence of denial' (Spender, 1990) is indeed powerful. Moreover, it is a silence compounded by our lack of tools – especially language – available to help us understand it. As Wilkinson (2005, p. 45) points out, in terms of studying human suffering,

> a sociological response to human suffering may require that we amplify unsettling questions of meaning and morality … and that we make abundantly clear, the terminal failure of understanding that takes place under the attempt to render the cultural grammar of suffering accountable to the rationality of scientific analysis.

Yet, as Sultana (1992, p. 19) reminds us, attending to the silences in ethnographic work can itself be a powerful tool that calls into question the epistemic and methodological status quo:

> Qualitative research, in itself, seems little concerned with the absences that frame (or are at the heart of) the narrative it weaves – the 'reality' it claims to reflect – or the process through which the phenomena are represented. The roots of ethnography are to be found deeply embedded in a regime of realism, inasmuch as it sets out to represent the empirical world that 'is.' There is, indeed, a danger that the ethnographic narrative entraps the writer and reader in this nominal positivist world; details of what happened, who spoke, what was said. It is the dictatorship of data … In these positivistic moments that govern ethnographic representation, silence has little value and can only be regarded negatively – an empty absence, which, in a matter of time, could and should become full of words.

Towards an inclusive post-humanist methodology

Despite these difficulties and frustrations, a dedicated – and growing – group of scholars remain committed to understanding human–animal relations for the most part because they want to contribute to an eradication of animal abuse and mistreatment. Many are sociologists precisely because a sociological approach is by definition attentive to power relations, and as a result, sociology is, in principle, in a strong position to address the mechanisms and structures that normalize animal oppression (Cudworth, 2016; Peggs, 2013; Taylor, 2011, Taylor and Sutton, 2018). However, a distinct sociological animal studies – one that 'raises questions about the exploitation and oppression of non-human animals' (Cudworth, 2016, p. 243) – remains marginal, with the majority of sociological work that does address other animals doing so from an anthropocentric and depoliticized stance (Taylor and Sutton, 2018). Similarly, while there have been calls for a sociology *for* other animals (Cudworth, 2016), 'there is a difference between acknowledging animal oppression and challenging it', which in turn has led to a call for a distinctly emancipatory animal sociology (Taylor and Sutton, 2018, p. 481). The reluctance of mainstream (and even some aspects of animal) sociology to adopt an emancipatory approach towards animal studies may well be the result of multiple factors, such as concerns over having one's work devalued and marginalized (e.g. Fraser and Taylor, 2016; Wilkie, 2015), or the fact that including other animals in a meaningful way requires a long hard look at the post-Enlightenment rationality that still pervades much of our discipline (Laue, 1989; Taylor, 2011). But it is also the case that the inclusion of other animals *per se*, particularly with the intention of advocating on their behalf, requires an overhaul of many of the methodological prescripts that sociologists hold dear. As feminist scholars pointed out regarding the study of women early in the development of feminist studies (Stanley and Wise, 1993), we can't simply tack animals onto humanist paradigms. Instead, those paradigms have to be changed if we want a truly *multispecies sociology*. Part of that change involves interrogating our research methods, given that our methods make the world (Law, 2004). If our methods exclude those about/for whom they purport to generate knowledge, then we are failing at the starting gate. And it is here that ethnography has radical potential.

However, while ethnography may have the potential to include other animals in ways that traditional methods do not, it is still rare to see animals included in research that purports to be about them and their relations with humans. The main reason may be that it is difficult to include animals in any ethnographic research unless it is visual. Ordinarily, they do not write or talk (in a language humans understand) so we cannot include them through our usual techniques. A secondary reason, which has much deeper roots and potential consequences, is likely that we simply did not think to include animals in ethnographic research because of long-held assumptions that animals do not matter or that they do not matter to those of us who study the social realm.

However, partly in recognition of its colonialist roots, ethnography itself is changing. As Skeggs (1999, p. 37) points out, this involves a shift in seeing and utilizing ethnography as a 'liberatory strategy. It is the shift from positioning others to authorizing others.' Melding this aspect of ethnography with insights gained from post-humanism, and the attendant legitimacy of investigating human–animal assemblages, does not just give us permission to look at and seek to understand and represent human–animal relationships but also enables us to use our scholarship to try to change them. One particularly promising development in this area is multispecies ethnography.

In acknowledging 'the human as a kind of corporeality that comes into being relative to multispecies assemblages, rather than as a biocultural given' (Ogden, Hall and Tanita, 2013, p. 6), multispecies ethnography opens the door to post-humanism for animal studies scholars but *only if* we give equal interest to humans and animals and we do not use the 'posthumanist turn' (Wolfe, 2010) or the 'animal turn' (Taylor, 2013) as a way to understand humanity. As it currently stands, much multispecies ethnography seems solely – or at least mainly – preoccupied with such anthropocentric concerns. As Ogden, Hall and Tanita (2013, p. 6) explain, much of the work done under the umbrella of multispecies ethnography to date 'has focused on the relations of multiple organisms (plants, viruses, human, and non-human animals), with a particular emphasis on understanding the human as emergent through these relations ("becoming")'. While seemingly innovative, in many ways, this is counterproductive, as it relegates multispecies ethnography to simply one more way to understand anthropocentric concerns, which effectively strips it of its radical potential and runs the risk of it becoming an intellectual fad.

While multispecies ethnography is grounded in work that may move it away from essentialist conceptions of 'the human' and/versus 'the animal' as discrete categories by keeping the focus on how *the human* emerges, anthropocentrism remains the organizing principle and concern. There is a clear need to divert multispecies ethnography away from this towards a consideration of its utility for a liberatory praxis. This has the advantage of returning it to both its ethnographic promise (of 'authorizing others') and its post-humanist roots, which here are read as attempts to 'decenter the human in ethics and theory' (Ogden, Hall and Tanita, 2013 p. 6).

Concluding thoughts: Multispecies ethnography and scholar-activism

We have argued that the new normal *demands* that we undertake research that advocates for other species (as well as the Earth and other humans) and that multispecies sociology is one way to achieve this if we accept that we need to disrupt conventional methodological prescripts. The strength of multispecies ethnography is that it points out the importance of other life and the interconnectedness of all beings, decentring the human. But we should not get carried away with the intellectualism of the project and forget how politics and power are present in any research – the very concerns that led us to it in the first place. As Sultana (1992, p. 26) notes,

> it is incorrect to pursue a strategy, such as descriptive ethnography, which, while satisfying, indeed gratifying, our 'need' to know, is largely ineffective in promoting transformation ... Radical ethnography can shed its subservience to the regimes of realism in order to become a meditative and reflexive vehicle.

That multispecies ethnography points to the symbiotic and cooperative nature of life is key. In many important ways, it feeds into and rests upon the antithesis of post-Enlightenment paradigms that stress competition and pure boundaries/binaries. As such, its political power is potentially revolutionary in an epistemic sense. But to realize this potential, those using multispecies ethnography must follow its internal logic. While the idea of human–animal (and other categories) 'assemblages' has liberatory potential in that it sets everything on a level playing field, as opposed to a hierarchical one, it also runs the risk of losing the animals in its portrayal of them in such a way. In other words, under such a formulation, they can be portrayed as objects just as easily as they can be portrayed as sentient beings. We need to be wary of this while at the same time making sure we are not reiterating essentialist calls of what 'the animal' – or indeed 'the human' – is. As currently configured, multispecies ethnography strays perilously close to losing the animal in such a way that human interests/politics – that is, its central concern is about becoming human – are prioritized.

Ogden, Hall and Tanita (2013, p. 16) state that their review of multispecies ethnography

> suggests the emergence of a very thoughtful MS political ecology, a kind of anti-essentialist approach that is mindful to the non-human in politics, though mainly politics in the classic sense of the term. We anticipate

a more critical engagement with how we approach 'politics' in the years to come.

We add our voices to this and extend it by pointing out that a more critical engagement has to include the politics of scholar-activism, whereby to be truly radical is to break with the idea of objectivity – beyond statements of individual subjectivity by the researcher – to an actual lived subjectivity that advocates on behalf of, in this case, non-human animals. Multispecies ethnography's intellectual heritage of post-humanism and its methodological reliance upon ethnography make it perfectly placed as

> ethnography is like a permanent ethical dilemma. It forces accountability and makes us assess the usefulness of theory and concepts … It is the only methodology that can show how complex processes are lived together and in contradiction. It should make us think about our relationships to others (not ourselves), aware that knowledge always bears the mark of its producer.
>
> (Skeggs, 1999, p. 48)

Throughout this chapter, we have argued that a crisis-ridden 'new normal' necessitates that we engage in praxis and that this mandates us to use sociological knowledge and tools to liberate other animals. We have argued that in order to do this we need to consider whether our research methods contribute to the anti-animal status quo or whether they help to challenge animal oppression. For us, part of this has been to examine our use of binaries. We have been concerned with how to complicate simplistic binaries while also recognizing how power and species privilege and oppression operate, often through such binaries. One way is to 'tell the stories that contact zones [between humans and other species] ignite' (Ogden, Hall and Tanita, 2013, p. 11). Yet, we need to do so carefully to ensure that we do not obviate/erase the realities of power discrepancies. Not everyone will agree with this idea; those who focus exclusively on humanity might well dismiss our metaphorical attempts to let the dogs (hens, sheep and birds) out of their cages. But for those of us who identify as scholar-activists and who want to ensure our politicized sociological scholarship contributes to the much-needed sense of urgency in the new normal, and thereby to actual change for animals, there is a clear need to use our sociological knowledge of how power and epistemology overlap to ensure we can reorient both post-humanism and multispecies ethnography to use them to their full emancipatory potential.

Notes

1 We realize that the terms are used to denote certain differences in the field, and that while anthrozoology, human–animal studies and animal studies may arguably be interchangeable, the activist and anti-capitalist stance of critical animal studies often denotes it as separate from the others. See Taylor and Twine (2014) for discussion.

2 We are not ignorant to the irony of supposedly needing to have our concerns legiti-
mated according to the standards set by the very forms of knowledge we are critiquing,
but in practical terms this needs to happen to open up spaces for these debates – for
example, having our work funded and accepted in peer-reviewed outlets.

References

Adams, C. (1990). *The Sexual Politics of Meat*. New York: Continuum.

Arluke, A. and Sanders, C. (1996). *Regarding Animals*. Philadelphia, PA: Temple
University Press.

Ascione, F.R. (1998). Battered women's reports of their partners' and their children's
cruelty to animals. *Journal of Emotional Abuse*, 1, 119–33.

Beck, A. and Katcher, A. (1996). *Between Pets and People: The Importance of Animal
Companionship*. West Lafayette, IN: Purdue University Press.

Becker, F. and French, L. (2004). Making the links: Child abuse, animal cruelty and
domestic violence. *Child Abuse Review*, 13, 399–414.

Becker, H. (1974). Photography and sociology. *Studies in the Anthropology of Visual
Communication*, 1, 3–26.

Best, S. (2009). The rise of critical animal studies: Putting theory into action and animal
liberation into higher education. *Journal for Critical Animal Studies*, 7(1), 9–53.

Blattner, C., Donaldson, S. and Wilcox, R. (2020). Animal agency in community: A
political multispecies ethnography of VINE sanctuary. *Politics & Animals*, 6. https://
journals.lub.lu.se/pa/issue/view/3116

Boddice, R. (ed.) (2011). *Anthropocentrism: Humans, Animals, Environments*. Leiden:
Brill.

Brozek, W. and Falkenberg, C. (2021). Industrial animal farming and zoonotic risk:
COVID-19 as a gateway to sustainable change? A scoping study. *Sustainability*,
13(16), 9251.

Canan, P. (1996). Bringing nature back in: The challenge of environmental sociology.
Sociological Inquiry, 66(1), 29–37.

Carrington, D. (2020, 27 April). Halt destruction of nature or suffer even worse
pandemics, say world's top scientists. *The Guardian*. www.theguardian.com/
world/2020/apr/27/halt-destruction-nature-worse-pandemics-top-scientists

Chagani, F. (2014). Critical political ecology and the seductions of posthumanism.
Journal of Political Ecology, 25, 424–36.

Cole, M. and Stewart. K (2014). *Our Children and Other Animals: The Cultural
Construction of Human–Animal Relations in Childhood*. Aldershot: Ashgate.

Cudworth, E. (2011). *Social Lives with Other Animals: Tales of Sex, Death and Love*.
Basingstoke: Palgrave Macmillan.

Cudworth, E. (2015). Killing animals: Sociology, species relations and institutionalized
violence. *The Sociological Review*, 63(1), 1–18.

Cudworth, E. (2016). A sociology for other animals: Analysis, advocacy, intervention.
International Journal of Sociology and Social Policy, 36(3/4), 242–57.

Dallimore, E.J. (2000). Feminist responses to issues of validity in research. *Women's
Studies in Communication*, 23(2), 157–81.

Davis, P. (2021, 3 December). COVID and climate change: A tale of two crises.
Newsroom NZ. www.newsroom.co.nz/peter-davis-a-tale-of-two-crises

DeMello, M. (2012) *Animals and Society: An Introduction to Human–Animal Studies*.
New York: Columbia University Press.

Donovan, J. (2018). Animal ethics, the new materialism and the question of subjectivity.
In A. Matsuoka and J. Sorenson (eds), *Critical Animal Studies: Towards Trans-species
Social Justice*. London: Rowman and Littlefield.

Drew, L. and Taylor, N. (2014). Engaged activist research: Challenging apolitical objectivity. In A. Nocella, J. Sorenson, K. Socha and A. Matsuoka (eds), *Defining Critical Animal Studies: An Intersectional Social Justice Approach for Liberation*. New York: Peter Lang.

Foley, D. (2002). Critical ethnography: The reflexive turn. *Qualitative Studies in Education*, 15(5), 469–90.

Foucault, M. (1991). What is an author? In P. Rabinow (ed), *The Foucault Reader*. New York: Pantheon.

Fox, N.J. and Alldred, P. (2020). Sustainability, feminist posthumanism and the unusual capacities of (post) humans. *Environmental Sociology*, 6(2), 121–31.

Fraiman, S. (2012). Pussy panic versus linking animals: Tracking gender in animal studies. *Critical Inquiry*, 39(1), 89–115

Fraser, H. and Taylor, N. (2016). *Neoliberalization, Universities and the Public Intellectual: Species: Gender and Class in the Production of Knowledge*. London: Palgrave Macmillan.

Fraser, H. and Taylor, N. (2020). Narrative feminist research interviewing with 'inconvenient groups' about sensitive topics: Affect, iteration and assemblages. *Qualitative Research*, 22(2), 220–35.

Gillespie, K. (2018). *The Cow with Ear Tag #1389*. Chicago: University of Chicago Press.

Gillespie, K. (2019). For a politicized multispecies ethnography: Reflections on a feminist geographic pedagogical experiment. *Politics & Animals*, 5. https://journals.lub.lu.se/pa/issue/view/2733

Gillespie, K. (2021). For multispecies ethnography. *EPE: Nature and Space*. https://doi.org/10.1177/25148486211052872

Hamilton, L. and Taylor, N. (2012). Ethnography in evolution: Adapting to the animal 'other' in organizations. *Journal of Organizational Ethnography*, 1(1), 43–51.

Hamilton, L. and Taylor, N. (2017). *Ethnography After Humanism: Power, Politics and Method in Multi-Species Research*. London: Palgrave Macmillan.

Harding, S. (1987). *Feminism and Methodology*. Bloomington, IN: Indiana University Press.

Harrison, S., Baker, M.G., Benschop, J., et al. (2020). One Health Aotearoa: A transdisciplinary initiative to improve human, animal and environmental health in New Zealand. *One Health Outlook*, 2, 4. https://doi.org/10.1186/s42522-020-0011-0

Hornborg, A. (2017). Artifacts have consequences, not agency: Toward a critical theory of global environmental history. *European Journal of Social Theory*, 20(1), 95–110.

Houston, M. and Kramarae, C. (1991). Speaking from silence: Methods of silencing and of resistance. *Discourse & Society*, 2(4), 387–99.

Kaijser, A. and Kronsell, A. (2014). Climate change through the lens of intersectionality. *Environmental Politics*, 23(3), 417–33.

Kemmerer, L. (2011). *Sister Species: Women, Animals and Social Justice*. Chicago, IL: University of Illinois Press.

Kochevar, D. and Werneck, G. (2021, 3 November). Preventing future pandemics starts with recognizing links between human and animal health. *The Conversation*. https://theconversation.com/preventing-future-pandemics-starts-with-recognizing-links-between-human-and-animal-health-167617

Kopnina, H. (2017). Beyond multispecies ethnography: Engaging with violence and animal rights in anthropology. *Critique of Anthropology*, 37(3), 333–57.

Latour, B. (2007). *Reassembling the Social: An Introduction to Actor-Network-Theory*. Oxford: Oxford University Press.

Laue, J. (1989). Sociology as advocacy: There are no neutrals. *Sociological Practice*, 7(1), Article 15. http://digitalcommons.wayne.edu/socprac/vol7/iss1/15

Law, J. (2004). *After Method: Mess and Social Theory*. London: Routledge.

Law, J. and Hassard, J. (1999). *Actor Network Theory and After*. Oxford: Wiley-Blackwell.

Law, J. and Mol. A. (2008) The actor enacted: Cumbrian sheep. In C. Knappett and L. Malafouris (eds), *Material Agency: Towards a Non-anthropocentric Approach*. New York: Springer.

MacGregor S. (2009). A stranger silence still: The need for feminist social research on climate change. *The Sociological Review*, 57(2 Suppl), 124–40.

MacGregor, S. (2014). Only resist: Feminist ecological citizenship and the post-politics of climate change. *Hypatia*, 29(3), 617–33.

McCann, H. (2016). Epistemology of the subject: Queer theory's challenge to feminist sociology. *Women's Studies Quarterly*, 44(3/4), 224–43.

Mills, C.W. (1959). *The Sociological Imagination*. Oxford: Oxford University Press.

Ogden, L., Hall, B. and Tanita, K. (2013). Animals, plants, people and things: A review of MSE. *Environment and Society: Advances in Research*, 4, 5–24.

Orzechowski, K. (2022). Global animal slaughter statistics and charts: 2022 update. *Faunalytics*. https://faunalytics.org/global-animal-slaughter-statistics-charts-2022-upd ate/?fbclid=IwAR02m22LYD8o4bD50JLL6p2k-BUj1hTeDSny2XFRrqovileOTl6U 9SycHVs

Peggs, K. (2013). The 'animal-advocacy agenda': Exploring sociology for non-human animals. *The Sociological Review*, 61, 591–606.

Pellow, D. and Brehm, H. (2013). An environmental sociology for the twenty-first century. *Annual Review of Sociology*, 39, 229–50.

Reddy, B. and Saier, M. (2020). The causal relationship between eating animals and viral epidemics. *Microbial Physiology*, 30, 2–8.

Riggs, D.W., Taylor, N., Signal, T., Fraser, H., and Donovan, C. (2018). People of diverse genders and/or sexualities and their animal companions: Experiences of family violence in a bi-national sample. *Journal of Family Issues*, 39, 4226–47.

Schally, J.L. (2017). *Legitimizing Corporate Harm: The Discourse of Contemporary Agribusiness*. Dordrecht: Springer.

Skeggs, B. (1999). Seeing differently: Ethnography and explanatory power. *Australian Educational Researcher*, 26(1), 33–53.

Smith, D. (1987). *The Everyday World as Problematic: A Feminist Sociology*. Atlanta, GA: Northeastern University Press.

Spender, D. (1990). Sounds of silence. *The American Voice*, 21, 106–11.

Stacey, J. and Thorne, B. (1985). The missing feminist revolution in sociology. *Social Problems*, 32(4), 301–16.

Stanley, L. (1990). *Feminist Praxis: Research, Theory and Epistemology in Feminist Sociology*. London: Routledge.

Stanley, L. and Wise, S. (1993). *Breaking Out Again: Feminist Ontology and Epistemology*. London: Routledge.

Strega, S. (2015). The view from the poststructural margins: Epistemology and methodology reconsidered. In L. Brown and S. Strega (eds), *Research as Resistance: Critical, Indigenous and Anti-oppressive Approaches*. Toronto: Canadian Scholars' Press.

Sultana, R. (1992). Ethnography and the politics of absence. *Qualitative Studies in Education*, 5(1), 19–27.

Sutton, Z. (2020). Researching towards a critically posthumanist future: On the political 'doing' of critical research for companion animal liberation. *International Journal of Sociology and Social Policy*, 41(3/4), 376–90.

Sutton, Z. and Taylor, N. (2019). Managing the borders: Static/dynamic nature and the 'management' of 'problem' species. *Parallax*, 25(4), 379–94.

Taylor, C. (1992). The politics of recognition. In A. Gutman (ed), *Multiculturalism: Examining the Politics of Recognition*. Princeton, NJ: Princeton University Press.

Taylor, C., Struthers Montford, K. and Kasprzycka, E. (2021). Introduction: Critical animal studies perspectives on COVID-19. *Animal Studies Journal*, 10(1), 1–6.

Taylor, N. (2011). Can sociology contribute to the emancipation of animals? In N. Taylor and T. Signal (eds), *Theorizing Animals: Re-thinking Humanimal Relations*. Leiden: Brill.

Taylor, N. (2013). *Humans, Animals and Society: An Introduction to Human–Animal Studies*. New York: Lantern Books.

Taylor, N. and Fraser, H. (2017) Slaughterhouses: The language of life, the discourse of death. In J. Maher, H. Pierpoint and P. Beirne (eds), *The Palgrave International Handbook of Animal Abuse Studies*. London: Palgrave Macmillan.

Taylor, N. and Fraser, H. (2019). Resisting sexism and speciesism in the social sciences: Using feminist, species-inclusive, visual methods to value the work of women and (other) animals. *Gender, Work & Organization*, 26(3), 343–57.

Taylor, N., Fraser, H., Stekelenburg, N. and King, J. (2022). Barbaric, feral or moral? Stereotypical dairy farmer and vegan discourses on the business of animal consumption. In L. Tallberg and L. Hamilton (eds), *The Oxford Handbook of Animal Organization Studies*. Oxford: Oxford University Press.

Taylor, N. and Sutton, Z. (2018). For an emancipatory animal sociology. *Journal of Sociology*, 54(4), 467–87.

Taylor, N. and Twine, R. (2014). *The Rise of Critical Animal Studies: From the Margin to the Centre*. London: Routledge.

Torres, B. (2007). *Making a Killing: The Political Economy of Animal Rights*. Chico, CA: AK Press.

Wilkie, R. (2015). Academic 'dirty work': Mapping scholarly labor in a tainted mixed-species field. *Society & Animals*, 23(3), 211–30.

Wilkinson, I. (2005). *Suffering: A Sociological Introduction*. Cambridge: Polity Press.

Willis, P. (1977) *Learning to Labour: Why Working-Class Kids get Working-Class Jobs*. London: Saxon House.

Wolfe, C. (2010). *What is Posthumanism?* Minneapolis, MN: University of Minnesota Press.

Wu, D. (2021, 10 November). Scott Morrison makes $1 billion pledge to find emerging technologies to reduce emissions and create jobs. *Sky News Australia*. www.skynews.com.au/australia-news/politics

Part II
Activism

4 (Not so) hidden barriers to a vegan-inclusive norm

The struggle against speciesism for compassionate children who could change the world

Lynda M. Korimboccus

Introduction

As outlined in the various chapters within this collection and elsewhere, non-human animals are subject to relentless exploitation and oppression by humans. In the time it may have taken you to read the opening sentence of this paragraph (say, five seconds), an estimated 42,700 non-human animals were killed worldwide for food: around 8,500 per second (Vegan Calculator, 2022). To reiterate, that is *eight and a half thousand deaths – every second.* For them, for humans and for the planet, this is unsustainable (Poore and Nemecek, 2018). Human veganism is a rational, reasonable response to such mistreatment; vegan advocacy is necessary to develop the move to a more sustainable and sympathetic way of living; and vegan sociologies provide useful theoretical bases for understanding the individual and institutional interactions that promote or prevent such social change. Today's children are tomorrow's consumers, and it is only once we understand how they are socialized into sustained speciesism – that is, believing non-human species to be less worthy than humans – that we can challenge this and undertake improvements that will ultimately benefit everyone. The 'sociological imagination' that C. Wright Mills (1959) encouraged us to employ more than 60 years ago outlined the need for us to think more systemically about society, to understand its complexities and to ultimately improve the lived experiences of those within it. Briefly applying some key sociological concepts of repression in social institutions and of individuals, this chapter will show how structural inequalities directly affect vegan children, indirectly affect the normalization of compassion and stall any subsequent move to a world free from harm. It should be noted that, throughout this chapter, systems and structures referred to are those of a progressively liberal modern Scotland or the wider United Kingdom. The additional barriers faced by vegans in more politically controlled places may be immeasurable, but nonetheless important. However, change should be possible, indeed welcome, in a country already committed to equality and diversity such as Scotland, whilst acknowledging the road ahead elsewhere may prove significantly more challenging.

DOI: 10.4324/9781003257912-7

Much has been written for many years in sociology about socioeconomic, gender-based, racial and other human inequalities (Gamoran, 2021) and ideas on how best to combat these have been the bread and (vegan) butter of sociologists for a long time (though progress to practically resolve these 'isms' is slow). Vegan sociology, though, is in its infancy, and takes a more holistic and intersectional approach to inequity, recognizing the interplay between human and non-human subjugation (Cherry, 2020; Sutton and Wrenn, 2020). As indoctrination into one's cultural values and norms begins in early childhood through the socialization process, it is important to recognize that those wielding social, political and economic power in a particular place are responsible for the folkways that follow. Introductory sociology classes often begin with a discussion of what 'society' actually means, and this initially vague definition leads to discussions of the spheres of influence within which we live. In most cases, these consist of the structures and institutions we encounter on a daily basis, such as the family, peers, education, employment, the media, religion, culture, politics, health and government – all of them reinforcing normative attitudes and associated behaviours, including speciesism and the associated derogation of the non-human. In response to these covertly oppressive systems, veganism becomes a subcultural curiosity of the few by the many (Christopher, Bartkowski and Haverda, 2018; Greenebaum, 2018).

Social structures

For children, indoctrination into normative speciesism effectively starts in the womb, where the dietary habits of their mother directly impact their development (Borge et al., 2017; *The Lancet*, 2018). Pregnant women in the United Kingdom (UK) are advised against consumption of various items, such as alcohol, certain raw eggs, liver, some dairy and some non-human animal meats including some species of fish (NHS, 2022) for fear of adverse effects on the foetus. Once born, most human babies in the UK are fed infant formula rather than breastfed (UNICEF UK, 2018), which is almost exclusively non-vegan, containing powdered animal milk and other animal-sourced additives such as lanolin (extracted from the wool of a sheep). A child's first walking shoes may be cow or calfskin, and many manufactured baby foods contain pieces of chicken, fish or other animal flesh. The most popular creams and potions for sore bottoms are developed by pharmaceutical companies, many of whom test their products or ingredients on animals (Johnson & Johnson, 2022). Children are often taken to the zoo, the aquarium or the city farm, where happiness at the sight of the animals is encouraged by caregivers. None of these activities is deemed unusual – indeed, they are normalized: promoted tourist attractions for families invariably include recommendations of locations that use animals as a secondary drawcard for visitors besides their main business, as well as those whose business relies primarily on incarcerated animals (Visit Scotland, 2022).

Many, if not most, of these same children will have one or more animals in their collection of toys and may in fact cherish a favourite cuddly rabbit or dog. Animal designs adorn drink bottles, bowls and spoons the world over, and favourite cartoon characters are often non-human animals. At the same time, those bowls frequently contain the body parts of dead animals, often the same species as their toys, and possibly the same species as their favourite character. For example, the pre-school series *Peppa Pig* and the *Babe* movie franchise are or were very popular with children, but most of those children will at the same time have been trained to accept the flesh of pigs in various guises as 'food', with the fate of farmed pigs simply overlooked. This is known as the Peppa Pig paradox (Korimboccus, 2020) and can happen just as easily with chickens or lambs as it can with pigs. Children's media, through books (Bowd, 1982; McCrindle and Odendaal, 1994), film (Bettany and Belk, 2011; Cole and Stewart, 2012, 2014; Hirschman and Sanders, 1997) and television (Korimboccus, 2021; Mills 2017; Paul 2015), reinforce many of these ideas, framing non-humans in ways so far removed from their natural existence that it is no wonder this disconnect exists between the representation of a species and the real animal. More than half of mainstream UK TV for young children contains animal characters in title or lead roles (Korimboccus, 2021), and children love them. This disconnect may be referred to in many ways: cognitive dissonance; wilful ignorance; the meat paradox or systematic aporia.

Despite recent evidence of a lack of speciesism in young children (aged nine to 11), and how similar they consider their non-human kin to fellow humans in a moral sense (McGuire, Palmer and Faber, 2022), they may, through the speciesist design of formal and informal education around them, remain oblivious to the reality of life for those species. Whether crafting with feathers, witnessing chick-hatching, engaging in egg-based bakery, studying 'farm' themes or eating school dinners, by the time connections are made, habits are well formed and more difficult to undo. These habits include the use of everyday language, the 'habitual patterns [of which] make meat-eating and factory farming seem natural' (Moore, 2014), their reality hidden.

Hidden barriers

Social class

The term 'hidden barriers' has in the past been reserved primarily for discussions of lack of opportunity in education or employment, and in education the *hidden curriculum* is a widely contested concept. Marxists see education as a system that disproportionately rewards individuals who already benefit from the existing class status of family through exposure to certain social, cultural and even linguistic experiences (Bourdieu and Passeron, 1990; Sullivan, 2001). They propose that the system is built around the passive conformity of the working-class majority through adherence to strict rules on timekeeping, uniformity and deference. Such a process is designed to disadvantage those in the lower echelons of society so that the ruling classes have a sufficient manual labour workforce to increase profits in an industrial, capital-driven society. They are introduced to authority and uniformity, prevented from meaningful participation and educated simply for their role as workers of the subject class (Banfield, 2015). They are simply a commodity within an industrial complex that seeks to maximize profits. That industrial complex includes those involved in (though often alienated from) the misery of non-human animals, whether in the disassembly, repackaging or representation of their body parts, skins or secretions. These are, sadly, more examples of cultural hegemony, played out in everyday capitalist society and protected by the powerful elite (Fitzgerald and Taylor, 2014).

The Weberian concept of elite self-recruitment proposed that those at the top of the industry decide the entry requirements for their peerage and exclude many who fail to make the grade (Mills, 1956; Pakulski, 2012; Weber, 1978). Even those able to rise through the ranks are likely to find themselves limited by their social start and less likely to achieve equity with their upper-middle-class peers once there (Crawford and Vignoles, 2014; The Sutton Trust, 2019). The elite tend to influence systemic decisions that continue to affect the institutions within which their own opportunities were created and taken. It is therefore understandable (although not excusable) for industry to seek to maintain the existing system that subliminally teaches children and young people that non-humans are a mere commodity from which profit is extracted. There is money to be made in murder. Children are taught through classroom activities, storybooks, television programmes and interactions with other parts of the social system that 'animals are *for*' this or that.

Sex and gender

For others, most notably some feminist theorists, the *hidden* or *gendered curriculum* exists as part of a patriarchal education system in which the vested interests of the most socially powerful (men) determine the nature of its design. In the United Kingdom at least, girls are no longer excluded from traditionally 'male' subjects, but nor are they encouraged to study them. Proactivity is vital to any balance redress. The discrimination against women spills over into employment and the notion that there exists a 'glass ceiling', preventing women from accessing top-level jobs and (although improving) boardroom chairs. Even as recently as 2021, there are only eight female CEOs in the top 250 UK companies (Vinnicombe et al., 2021), with many companies still underachieving their target of one-third women on corporate committees. The system subtly prevents such promotion to or through the ranks in line with male counterparts. The expectations of even a twenty-first-century woman's life can make it difficult to manage and maintain a healthy work–life balance.

Non-human elements of nature are often viewed in terms of the feminine – for example, 'Mother Earth' – with many European languages of Latin origin using gender-based pronouns. However, whereas the English language settles for a more universal 'the', it is often paired with an adjective to create derisory, animal-based expressions (Goatly, 2006) to suggest weakness or emotionally driven illogicality. Decades ago, ecofeminists such as Carol J. Adams (1991) and Greta Gaard (1993) recognized the cultural conflation of women with nature, both rendered a resource for men and both experiencing endless exploitation. Vegan feminism highlights the use of the body in particular by those in power: whether for eggs, milk, offspring or gratification, the utility of the human and non-human female body for commercial gain is widespread and ongoing (e.g. Adams, 1991; Wrenn, 2017). The normalization of such attitudes and practices leaves a young female vegan child with more than one mountain to climb in being accepted as potentially strong and successful, individual and independent, with beliefs in equality and justice being overlooked.

Philosophical belief

Regularly making British news headlines are cases of sectarian segregation or stigma. It is widely accepted at this point in the twenty-first century that one's belief systems should be embraced as part of a multicultural society, and the UK *Equality Act 2010* reflects this. Recently joining the ranks of beliefs by which an individual should be treated equally is veganism. As previously mentioned, this philosophy extends to all areas of life: from toothpaste and other toiletries used in the morning to clothing and footwear choices for the day; from dinner plate contents of an evening to the household products used to clean them; from language use about other species to wider ethical choices of a television programme, family outing, charitable giving and even potential employer.

A recent UK employment law case brought by vegan Jordi Casamitjana against his former employers resulted in veganism being acknowledged as a protected characteristic under equality law (HM Courts & Tribunals Service, 2020), as it is a belief system requiring strict adherence to a nonviolent, antispeciesist way of life. Veganism addresses not only the inequalities mentioned above but also the fundamental difficulty with the commodification of our society, including the rights of women, low-paid workers, Indigenous peoples and more besides, as 'one struggle' (Jenkins and Stanescu, 2014, p. 74). According to the UN Convention on the Rights of the Child (UNCRC) (UNICEF UK 1990), children have the right to be heard, to express themselves freely and to receive reliable information, among other rights and responsibilities. Yet, vegan children are often silenced within a speciesist system that fails to recognize their values and beliefs equally with those of others.

Interactions and intersections

Interactionists propose that how we interpret communications within and between relationships is responsible for norm maintenance: that the perception of individuals can directly alter their experiences of institutions such as education and those within them (Blumer, 1969). This includes how one might understand everyday symbols such as language (including non-verbal communication) and the complex formation of personal and social identities. As well as Erving Goffman's (1956) work on how to manage one's identity when it is a stigmatized one (see also Greenebaum, 2018), Becker's (1963) 'Outsiders' study made clear the danger of labelling one another according to arbitrary cultural categories and allowing those labels to consume individuals, rendering their identities 'deviant'. Stereotypes of vegans as pale, malnourished, party-poopers persist (Twine, 2014), although this is now regularly and publicly countered by healthy, sporty vegan role models such as Formula 1 champion Lewis Hamilton, boxer David Haye and tennis player Venus Williams (Gabardi, 2019; *The Telegraph*, 2017). Add to these well-known celebrities such as Lizzo and Fat Gay Vegan, who are part of an important movement highlighting the expression of individuality, challenging multiple norms and mores of society along the way (Ritschel, 2020). Stanley Cohen's (1972) study of Mods and Rockers in late-1960s England illustrated just how powerful (but also damaging) labelling and stereotypes can be, particularly when exacerbated by mass media hyperbole. More than 50 years on, this is no less evident.

Charles Cooley's (1902) idea of the reflected self is also pertinent here. Cooley's view is that our behaviour is directly linked to our response to how we feel others view us, so real or imagined negative interactions are more likely to produce a negative outcome. Belonging to any marginalized group experiencing prejudice, then, could actively prevent the sense of belonging so sought after, as children navigate and negotiate everyday exchanges with others. The vegan child in the classroom may experience 'outsider' treatment in the form of humiliation, hostility or exclusion, despite veganism being an ethic that seeks to be cruelty free and fully inclusive (Stewart and Cole, 2018). If the child also happens to belong to one or more additional minority groups, then the only way to even start to understand this experience is through the lens of intersectional veganism.

Media representation and responsibility

As alluded to in the brief discussion of Cohen's (1972) study, it is widely acknowledged that the mass media have a significant influence in terms of beliefs, behaviours and attitudes. In particular, television programming aimed at children carries a huge responsibility in terms of its impact (Strasburger, 2004). This effect is clear in movies and television, where non-human characters are favourites of children the world over but often mere stereotypes of different species (Korimboccus, 2020, 2021). In gaming, it could be said that, as the sexualization of the early character of Lara Croft in *Tomb Raider* may have been held partially responsible for ongoing sexism (Engelbrecht, 2020), the anthropomorphism of non-human characters such as the Teenage Mutant Ninja Turtles may contribute to the persistence of speciesism (Simons, 2002).

Many superheroes on both the large and small screens have origins in animal species – Wolverine, Spiderman, Batman and other DC Comics characters, as well as Ladybug, Cat Noir and other Miraculous characters. Scripts allow characters to show off their superpowers, often exaggerated versions of the species' natural abilities – for example, the web-spinning powers of Spiderman. He remains a favourite of many children (and adults), although in some places wariness of spiders is understandable – indeed, one episode of the UK-created *Peppa Pig* was banned from Australian TV for describing spiders as harmless, given that some spider species found in Australia (both indoors and out) can be lethal to humans (Zhou, 2017). Nonetheless, many a truly harmless UK-resident spider may have met their demise on the underside of a Scottish teacher's classroom stapler.

Children and animal advocacy

Children are encouraged to feel affection for animals through various classroom and other educational activities, yet often these activities are unwittingly speciesist in nature, so ingrained is speciesism in the thoughts and actions of most of society. Child–animal relationships are believed to be beneficial, although more can be done to encourage this in a more compassionate way (Pattnaik, 2004). Unfortunately, too many experiences are offered by those who make a profit from the incarceration of non-humans (zoos or farms, for example) and although 'humane education' is a growing learning and teaching area, it is one that should be incorporated throughout a curriculum, rather than as a stand-alone topic, or tacked on to a lesson plan as an afterthought. Pattnaik (2004, p. 100) suggests activities within science, art and history, among other resources in literature to collectively 'connect children to an inner self that is inherently kind, compassionate, empathic, and justice-oriented', making them more open and ultimately improving society for everyone. This was the driving force behind Mills' (1956) 'sociological imagination'; yet, it is often forgotten by so many studying the discipline of sociology – or indeed teaching.

Conclusion

Whether dealing with 'glass', 'class' (Friedman, Laurison and Miles, 2015) or even 'grass' ceilings, inequalities unfortunately remain prevalent – albeit in the case of plant-based pupils, relatively fresh to the table for consideration. It is the job of educators and caregivers to first acknowledge the moral considerations of vegan children and incorporate these ethics as a lesson for all in their care (Korimboccus, 2023; McGuire, Palmer and Faber, 2022). Vegan children are subject to stigmatization in similar ways to other minority groups – often subtle, but there nonetheless (Cole and Morgan, 2011; Twine, 2014). Where there is injustice, though, there is an opportunity to redress the imbalance. That speciesism hides behind barriers of normativity renders it no less unacceptable, especially when it creates unequal treatment towards compassionate, cruelty-free children. It has long been argued that ignorance is no defence in law nor can it be accepted as a defence in any institution responsible for influencing the next generation. These children could change the world – let us help them to do so!

References

Adams, C.J. (1991). Ecofeminism and the eating of animals. *Hypatia*, 6(1), 125–45.

Banfield, G. (2015). Marx and education: Working with the revolutionary educator. *Journal for Critical Education Policy Studies*, 13(3), 8–28.

Becker, H. (1963). *Outsiders: Studies in the Sociology of Deviance*. London: Free Press.

Bettany, S. and Belk, R.W. (2011). Disney discourses of self and other: Animality, primitivity, modernity, and postmodernity. *Consumption Markets & Culture*, 14(2), 163–76.

Blumer, H. (1969). *Symbolic Interactionism: Perspective and Method*. Englewood Cliffs, NJ: Prentice Hall.

Borge, T.C., Aase, H., Brantsaeter, A.L. and Biele, G. 2017. The importance of maternal diet quality during pregnancy on cognitive and behavioural outcomes in children: A systematic review and meta-analysis. *BMJ Open*, 7(9), e016777.

Bourdieu, P. and Passeron, J. (1990). *Reproduction in Education, Society and Culture*. London: Sage.

Bowd, A.D. (1982). Young children's beliefs about animals. *The Journal of Psychology*, 110(2), 263–66.

Cherry, E. (2020). Emancipatory vegan sociology: Where are we going? Where have we been? 'Worldly Togetherness?' inaugural conference. *International Association of Vegan Sociologists*. https://youtu.be/Mf9LHBMgkeE

Christopher, A., Bartkowski, J. and Haverda, M. (2018). Portraits of veganism: A comparative discourse analysis of a second-order subculture. *Societies*, 8(5), 55–75.

Cohen, S. (1972). *Folk Devils and Moral Panics*. London: MacGibbon & Kee.

Cole, M. and Morgan, K. (2011). Vegaphobia: Derogatory discourses of veganism and the reproduction of speciesism in UK national newspapers. *British Journal of Sociology*, 61(1), 134–53.

Cole, M. and Stewart, K. (2012). *Puss in Boots* (2011) DreamWorks Animation, 90 min. Reviewed by Matthew Cole and Kate Stewart. *Journal for Critical Animal Studies*, 10(1), 182–92.

Cole, M. and Stewart, K. (2014). *Our Children and Other Animals: The Cultural Construction of Human–Animal Interaction in Childhood*. Aldershot: Ashgate.

Cooley, C.H. (1902). *Human Nature and the Social Order*. New York: Charles Scribner's Sons.

Crawford, C. and Vignoles, A. 2014. Heterogeneity in Graduate Earnings by Socio-economic Background: IFS Working Paper W14/30. https://ifs.org.uk/uploads/publications/wps/WP201430.pdf

Engelbrecht, J. (2020). The new Lara phenomenon: A postfeminist analysis of *Rise of the Tomb Raider*. *Game Studies*, 20(3). http://gamestudies.org/2003/articles/engelbrecht

Fitzgerald, A.J. and Taylor, N. (2014). The cultural hegemony of meat and the animal industrial complex. In N. Taylor and R. Twine (eds), *The Rise of Critical Animal Studies: From the Margins to the Centre*. London: Routledge.

Friedman, S., Laurison, D. and Miles, A. (2015). Breaking the 'class' ceiling? Social mobility into Britain's elite occupations. *The Sociological Review*, 63(2), 259–89.

Gaard, G. (1993). *Ecofeminism: Women, Animals, Nature*. Philadelphia, PA: Temple University Press.

Gabardi, C.S. (2019). 15 Elite Vegan Athletes Who Prove You Don't Need Meat. *Eluxe Magazine*. https://eluxemagazine.com/people/elite-vegan-athletes

Gamoran, A. (2021). Sociology's role in responding to inequality: Introduction to the special collection. *Socius*, 7. https://doi.org/10.1177/23780231211020201

Goatly, A. (2006). Humans, animals, and metaphors. *Society & Animals*, 14(1), 15–37.

Goffman, E. (1956). *The Presentation of Self in Everyday Life*. New York: Anchor Books.

Greenebaum, J. (2018). Vegans of color: Managing visible and invisible stigmas. *Food, Culture & Society*, 21(5), 680–97.

Hirschman, E.C. and Sanders, C.R. (1997). Motion pictures as metaphoric consumption: How animal narratives teach us to be human. *Semiotica*, 115(1/2), 53–79.

HM (Her Majesty's) Courts & Tribunals Service. (2020). Mr J Casamitjana Costa v The League Against Cruel Sports: 3331129/2018, Employment Tribunal decision. https://assets.publishing.service.gov.uk/media/5e3419ece5274a08dc828fdd/Mr_J_Casamitjana_Costa_v_The_League_Against_Cruel_Sports_-_3331129-18_-_Open_Preliminary_Hearing_Judgment___Reasons.pdf

Jenkins, S. and Stanescu, V. (2014). One struggle. In A.J. Nocella, J. Sorenson, K. Socha and A. Matsuoka (eds), *Defining Critical Animal Studies: An Intersectional Social Justice Approach for Liberation*. New York: Peter Lang.

Johnson & Johnson (2022). Humane care and use of animals policy. www.jnj.com/about-jnj/policies-and-positions/humane-care-and-use-of-animals-policy

Korimboccus, L.M. (2020). Pig ignorant: The Peppa Pig paradox – investigating contradictory childhood consumption. *Journal for Critical Animal Studies*, 17(5), 3–33.

Korimboccus, L.M. (2021). Animal representation on UK children's television. *Climate, Creatures and COVID-19: Environment and Animals in Twenty-First Century Media Discourse: A Networking Knowledge Special Issue. Networking Knowledge: Journal of the MeCCSA Postgraduate Network*, 14(2), 41–64.

Korimboccus, L.M. (2023). Friends, not food: How inclusive is education for young vegans in Scotland? In G. Lalli, A. Turner and M. Rutland (eds), *Food Futures in Education: Policy, Curricula and Society*. London: Routledge & CRC Press.

McCrindle, C.M.E. and Odendaal, J.S.J. (1994). Animals in books used for preschool children. *Anthrozoös*, 7(2), 135–46.

McGuire, L., Palmer, S.B. and Faber, N.S. (2022). The development of speciesism: Age-related differences in the moral view of animals. *Social Psychological and Personality Science*. https://doi.org/10.1177/19485506221086182

Mills, B. (2017). *Animals on Television*. London: Palgrave Macmillan.

Mills, C.W. (1956). *The Power Elite*. Oxford: Oxford University Press.

Mills, C.W. (1959). *The Sociological Imagination*. Oxford: Oxford University Press.

Moore, A.R. (2014). That could be me: Identity and identification in discourses about food, meat and animal welfare. *Linguistics & the Human Sciences*, 9(1), 59–93.

NHS (National Health Service) Inform (2022). Eating well in pregnancy. www.nhsinform.scot/ready-steady-baby/pregnancy/looking-after-yourself-and-your-baby/eating-well-in-pregnancy

Pakulski, J. (2012). The Weberian foundations of modern elite theory and democratic elitism. *Historical Social Research*, 37(1), 38–56.

Pattnaik, J. (2004). On behalf of their animal friends: Involving children in animal advocacy. *Childhood Education*, 81(2), 95–100.

Paul, E.S. (2015). The representation of animals on children's television. *Anthrozoös*, 9(4), 169–81.

Poore, J. and Nemecek, T. (2018). Reducing food's environmental impacts through producers and consumers. *Science* 360(3692), 987–92.

Ritschel, C. (2020, 30 June). Lizzo shares what she eats in a day as a new vegan. *Independent*. www.independent.co.uk/life-style/lizzo-vegan-tiktok-meals-breakfast-lunch-plant-based-a9594571.html

Simons, J. (2002). *Animals, Literature and the Politics of Representation*. London: Palgrave Macmillan.

Stewart, K. and Cole, M. (2018, 1 November). Vegans: Why they inspire fear and loathing among meat eaters. *The Conversation*. https://theconversation.com/vegans-why-they-inspire-fear-and-loathing-among-meat-eaters-106015

Strasburger, V.C. (2004). Children, adolescents and the media. *Current Problems in Pediatric Adolescent Health Care*, 34, 54–113.

Sullivan, A. (2001). Cultural capital and educational attainment. *Sociology*, 35(4), 893–912.

Sutton, Z. and Wrenn, C.L. (2020, 2 August). Vegan sociology – what and why? *Freedom of Species*. Podcast. www.3cr.org.au/freedomofspecies/episode-202008021300/vegan-sociology-what-and-why

The Equality Act. (2010). London: The Stationery Office. www.legislation.gov.uk

The Lancet. 2018. *Preconception Health*. Series from *The Lancet* journals. www.thelancet.com/series/preconception-health

The Sutton Trust. (2019). *Elites in the UK: Pulling Away? Social Mobility, Geographic Mobility and Elite Occupations*. London: The Sutton Trust.

The Telegraph (2017, 4 January). Green machines: The superstar athletes you never knew were vegan. *The Telegraph*. www.telegraph.co.uk/health-fitness/body/green-machines-the-superstar-athletes-you-never-knew-were-vegan

The Vegan Calculator. (2022). Animal slaughter counter. https://thevegancalculator.com/animal-slaughter

Twine, R. (2014). Vegan killjoys at the table: Contesting happiness and negotiating relationships with food practices. *Societies*, 4(4), 623–39.

UNICEF UK. (1990). *The United Nations Convention on the Rights of the Child*. London: UNICEF.

UNICEF UK. (2018). Breastfeeding in the UK. www.unicef.org.uk/babyfriendly/about/breastfeeding-in-the-uk

Vinnicombe, S., de Largy, C., Tessaro, M., Battista, V. and Anderson, D. (2021). *The Female FTSE Board Report 2021*. Cranfield: Cranfield University.

Visit Scotland. (2022). Scotland's unmissable family attractions. www.visitscotland.com/holidays-breaks/family/unmissable-attractions

Weber, M. (1978 [1922]). *Economy and Society: An Outline of Interpretive Sociology*. Berkeley, CA: University of California Press.

Wrenn, C.L. (2017). Towards a vegan feminist theory of the state. In D. Nibert (ed), *Animal Oppression and Capitalism*. Santa Barbara, CA: Praeger.

Zhou, N. (2017, 5 September). Peppa Pig 'spiders can't hurt you' episode pulled off air in Australia – again. *The Guardian*. www.theguardian.com/tv-and-radio/2017/sep/05/peppa-pig-spiders-cant-hurt-you-episode-pulled-off-air-in-australia-again

5 Incorporating a structural approach into animal advocacy

Nick Pendergrast

Introduction

In the current 'new normal' state of being as a result of the COVID-19 pandemic, as well as the escalating climate crisis, significant shifts in society have already occurred. The pandemic has shown that dramatic changes to society are possible and has led many people to re-evaluate their individual lives and their communities, including city design and public spaces, workplaces and much more (Matthews, 2020b; Plimmer, 2020; Thorpe, 2020). In this context, the environment is ripe for critically investigating food systems and our relationships with other animals more generally. This is particularly important when the consumption of animals has contributed significantly to climate change, through the environmental impact of animal agriculture, with vegan diets having a much lower climate impact than diets including animal products (e.g. Bruno et al. 2019; Clark, Hill and Tilman, 2018; Martin and Brandão 2017). It is also very relevant in light of the COVID pandemic, due to the zoonotic origins of COVID-19 (Tiwari et al., 2020; Ye et al., 2020). There are already signs of a growing interest in eating plant-based foods, with a 1,320 per cent increase in plant-based 'meats' being featured on the menus of restaurants in the United States compared with the start of the pandemic (Webber, 2021). The growing interest in plant-based food has been driven predominantly by health concerns, although environmental issues also play a key role (Webber, 2021). These individual transformations are covered in Chapter 6 of this book.

The animal advocacy movement often focuses on encouraging individuals to make consumer changes on behalf of animals, such as rejecting the consumption of all animal-derived products through veganism (Wrenn, 2018). Such advocacy emphasizes individual agency and the capacity to reject dominant attitudes and practices towards animals. While such individual changes are important, sociology also draws attention to broader structures that influence society and shape individual choices (Cudworth, 2005). This chapter will review the existing literature, applying a sociological analysis to animal advocacy (e.g. Cherry, 2021; Donaldson and Kymlicka, 2015; Wrenn, 2018) and will build on this literature by making more direct connections to *how* these ideas can be incorporated into animal activism. With a lot of activism being put 'on

DOI: 10.4324/9781003257912-8

hold' during the pandemic, this is the perfect time for activists to reflect on and re-evaluate the effectiveness of their campaign messaging.

This chapter provides a framing analysis of campaigns by two organizations based in Melbourne, Australia. It focuses on environmental advocacy by Vegan Rising and an anti-dairy campaign by Melbourne Cow Save. The framing analysis will assist in bridging the connection between academic work on a structural approach towards animal advocacy and activist campaigns. Frames shape how audiences understand and interpret issues and events, as well as their evaluations of political action (Carragee and Roefs, 2004; Kruse, 2001). Framing is based on 'principles of selection, emphasis, and presentation composed of little tacit theories about what exists, what happens, and what matters' (Gitlin, cited in Freeman 2013, p. 95). Once again, this chapter will build on the existing literature on the framing of animal advocacy campaigns (e.g. Freeman, 2010; Jasper and Poulsen, 1995; Williams, 2012), not only by analysing the campaigns but also by more directly intervening and suggesting alternative framing that could incorporate the structural issues identified above.

The analysis provided here is an example of research that aims 'for closer engagement between academics and the social movements that they study' to provide constructive and sympathetic critiques of these movements to advocate for more effective strategies (Matthews, 2020a, p. 591). Applying such an approach in the context of vegan advocacy is a part of the emerging field of vegan sociology, which 'seeks to further understanding of veganism and the factors that enable more effective pursuit and perpetuation of vegan lifestyles and messaging' (International Association of Vegan Sociologists, 2021). I took part in both of the actions discussed as an activist, which is consistent with a scholar-activist approach (for examples of researchers incorporating their activist experiences into their writing, see Matthews, 2020a; Scheper-Hughes, 1995; Smith and Glidden, 2012). As a result, these examples both come from an Australian context, although throughout the chapter examples from other countries are also given – mainly from the Western world.

Background

The structure–agency debate within sociology

The structure–agency debate is a central one within sociology (Cudworth, 2005). Agency is the ability to make our own choices, including those that go against the structures of society, which can include institutions such as the media and advertising, as well as social norms such as gender norms. While acknowledging individual agency, sociologists have a strong focus on the impact of social structures in shaping individual choices. If an individual lived in a different society or even under different circumstances within the same society – such as being a different gender or class – they may well have very different tastes and preferences (Baran, 2017; Germov and Hornosty, 2009;

Germov and Poole, 2015). Various sociological theories place different degrees of emphasis on structure and agency, but while it is useful to think about structure and agency as two opposites as a starting point for thinking about the concepts, a more developed analysis views structure and agency as interrelated (Cudworth, 2005). It is important to see the connection between the two concepts and acknowledge that 'structure is the mechanism through which action takes places and at the same time, agency produces and alters structure' (Giddens, cited in Cudworth, 2005, p. 61). This analysis has significant overlap with the central sociological concept of sociological imagination, which focuses on the connections between personal troubles and broader public issues (Mills, 1959).

These concepts of structure and agency are useful for understanding an endless amount of issues. Maneesha Deckha (2008) and more recently Stephanie Baran (2017) both analyse the campaigns of People for the Ethical Treatment of Animals (PETA) in a manner that incorporates these concepts. The focus is specifically on campaigns involving sexualized representations of women, highlighting the individual agency of the women taking part in the campaigns but also the broader structural implications of these campaigns. Similarly, anthropologist Nicole Constable (2009, p. 57) looks into ethnographic research on migrant women to challenge the 'agent–victim binary', arguing that it is important to avoid 'depicting women migrants as either passive victims who lack the ability to make choices or active agents who have full control of their circumstances'. These concepts are also very relevant for understanding who is responsible for animal exploitation and finding ways to challenge it.

Structure, agency and animal exploitation

An analysis that focuses purely on individual responsibility for animal exploitation, which is common within the animal advocacy movement, is limited in that it ignores the structural factors and institutions that also contribute to harm to animals, including the role of animal-exploiting industries and governments (Wrenn, 2018). Another important structural constraint is the dominance of the social norm of speciesism (discrimination against other animals) and 'the hegemonic meat and dairy food culture' (Howard, 2021, p. 4) that views consuming animals as legitimate and 'normal' (Cherry, 2021). Historian Gonzalo Villanueva (2018, p. 236) acknowledged such barriers when discussing veganism as an example of lifestyle activism, arguing that, 'Enormous social obstacles existed for practitioners of lifestyle politics, such as contending with entrenched norms and values and even alienation from family or friends'. Sue Donaldson and Will Kymlicka (2015) raise the issue that support from communities and institutions, which is really important in maintaining the practice of veganism, is not always present, meaning that in relying primarily on individual transformation through promoting individual veganism, the movement faces an uphill battle. While the growing interest in plant-based eating discussed above means that veganism does not depart from social norms as much as it did in the past (Webber, 2021), cultural barriers remain in place.

In a study on younger people and their willingness to choose meat alternatives, just under half of the respondents viewed meat alternatives as normal, with this being a significant barrier to consuming them (Bogueva and Schmidinger, 2021). One respondent cited these cultural barriers as a reason to avoid meat alternatives, arguing in terms of social norms:

> [Meat alternatives are] not normal. The normality of eating meat alternatives I believe comes with the societal acceptance and at the moment this is not the norm. I prefer to be part of the society and to follow its norms.
>
> (cited in Bogueva and Schmidinger, 2021, p. 28)

Such a blatant admission of desiring to follow norms is perhaps rare. However, while humans are not simply 'puppets on strings' and do have agency, we often unconsciously adopt the norms of our society when reaching what are sometimes portrayed as totally independent decisions outside of a sociological context (Germov and Poole, 2015, pp. 5–7), including decisions around food choices (Howard, 2021).

Sociologist Corey Wrenn's (2018) article 'How to Help When It Hurts? Think Systemic' makes an important contribution to this topic. While the article focuses on the very specific issue of the moral dilemma of what to feed rescued carnivorous animals, it has much broader application. Wrenn's (2018) article is very important for powerfully challenging the 'supply and demand' model so dominant in vegan discourse. Arguments from within the vegan movement often emphasize the role of individual consumers in 'voting with their dollar' and creating a demand for animal products or alternatively plant-based products. However, Wrenn's (2018, p. 152) analysis argues that such an approach is overly simplistic; instead, she highlights the role of producers and governments, making the case 'that food systems are seller-controlled rather than buyer-controlled'. Wrenn (2018) places the blame for the speciesist food system on corporate and government elites rather than individual consumers, arguing that consumer power is minimal within this system. She provides a number of examples to support this argument about corporations and governments primarily being responsible for animal consumption, citing advertising from animal-using industries, favourable government policies towards animal agriculture, state nutritional advice promoting animal consumption and 'ag gag' laws being introduced by governments to attempt to reduce activist efforts to challenge animal agriculture. Overall, Wrenn (2018, p. 172) makes a powerful case that, '"Meat" and dairy production is artificially high, and is forced into the food system regardless of consumer desires'. As a result, she points out that 'consumer choice is not especially well-suited to dismantling the speciesist system given the governmental and industrial control', framing veganism as 'important only as a political protest' rather than as an action that will directly reduce the demand for animal exploitation (Wrenn, 2018, p. 166). Similarly, others have argued for veganism in terms of prefigurative politics, of living out the world you want to see in the future and attempting to put it into practice now, rather than using a supply and demand analysis (Cherry, 2021; Petray and Pendergrast, 2018).

Wrenn's analysis is consistent with a structural, Marxist approach, with Moishe Postone (1993, p. 184), in a book exploring Marx's theory, arguing that a Marxist perspective highlights that under capitalism, 'on a deep, systemic level, production is not for the sake of consumption'. Donaldson and Kymlicka (2015, pp. 53–54) argue along similar lines in highlighting the limitations of vegan outreach (animal activists encouraging individual members of the public to become vegan):

> The implicit model of vegan outreach is a uni-directional arrow: you act on individual conscience, and eventually there are enough conscientious individuals to magically transform institutions. In reality, however, institutions are constantly acting upon individuals, undermining, frustrating, and co-opting individual efforts and desires. These political and institutional structures must be the direct focus of AR advocacy and organization.

This chapter builds on the work of theorists such as Donaldson and Kymlicka (2015), as well as Wrenn (2018), by looking more directly into exactly *how* animal activists can reframe their messaging to incorporate these ideas and engage more structurally. However, as discussed above, a sociological analysis should also incorporate an acknowledgement of individual agency. In this case, this involves the ability of individuals to avoid the consumption of animal products despite their widespread availability, promotion and acceptance within society. Returning to structures, this ability will be greater for some than for others, depending on their context and position within society, including factors such as geographical location and class (Cherry, 2021; Harper, 2010; Howard, 2021; Wrenn, 2011). Nevertheless, it is important to acknowledge that structures are not 'set in stone', but change over time, partly as a result of individual actions (Cudworth, 2005). Vegan outreach, while limited for the reasons specified above, should absolutely be a *part* of the animal advocacy movement. The purpose of this chapter is not to reject vegan outreach, individual veganism or individual change generally, but rather to question its dominance within animal advocacy and suggest ways in which animal advocates can include a greater acknowledgement of the structural factors responsible for animal exploitation, including in vegan outreach.

It is also important to note the growing number of existing examples of animal advocates engaging at a more structural level. An important example in the United Kingdom is Animal Rebellion, which aims to create a mass movement to achieve a plant-based food *system* (Animal Rebellion, 2021c). This organization has also focused on the *production* of animal products, rather than the typical focus of vegan outreach at the consumption/demand end of the process. This has included blockading Arla Foods, one of the largest manufacturers of dairy in the world, and fast food giant McDonald's, urging both of them to transition to being totally plant-based by 2025 (Animal Rebellion, 2021a, 2021b). Activists in Australia have also blockaded slaughter-houses as part of the Dominion Animal Liberation disruption, an action that was timed to coincide with the one-year anniversary of the documentary *Dominion*, which sheds light on animal exploitation in Australia (Hope, Webb and Bourke, 2019). This targeting of animal-using industries is nothing new – the illegal direct action movement the Animal Liberation Front engaged in a wide range of more militant tactics targeting animal industries, most commonly vandalism but also occasionally freeing animals and arson, with these activities peaking in the late 1990s in the United States (Glasser, 2011).

In Canada, Nation Rising (2021a) focuses on political advocacy, 'lobbying to shift multi-billion dollar government subsidies away from animal agriculture and towards the creation of a plant-based food system'. The group's recent advocacy has included the open letter 'Transition Canada to Plant-based Food System to Prevent Future Pandemics' to Canadian government ministers. As the title suggests, the focus is predominantly on human health, although environmental concerns are also raised (Nation Rising, 2021b). A similar campaign on a global scale is the Plant-Based Treaty by the Climate Save Movement. This international treaty is a companion to the UN Framework Convention on Climate Change/Paris Agreement and aims to halt the environmental harm caused by animal agriculture. Similar to the messaging of the organizations discussed above, the group calls for, 'An active transition away from animal-based food systems to plant-based food systems' (Climate Save Movement, 2021). This treaty has been endorsed by 17 politicians from a range of political parties in the United Kingdom (UK Parliament, 2021), as well as all three politicians from the Animal Justice Party in Australia (Laurence, 2021). This Australian political party also has policies of ending subsidies to animal-using industries and introducing a tax on animal products (Animal Justice Party, 2017), which are both systemic responses to animal exploitation. The tax specifically is an attempt to incentivize the consumption of plant-based products, rather than the usual approach of encouraging individuals to consume plant-based products within a system that promotes animal products, as discussed earlier. Finally, another attempt to tackle animal exploitation at the production end is the Rancher Advocacy Program by the US-based Rowdy Girl Sanctuary, which assists those working in animal agriculture to transition to plant-based agriculture (Smith, 2019). The following framing analysis will draw inspiration from these recent examples within the animal advocacy movement, as well as from other social movements and the structure–agency debate within sociology discussed above.

Framing and animal advocacy

Framing

Frames set parameters for 'what is going on' (Bateson, cited in Oliver and Johnston, 2000, p. 5) – different frames present different ways in which to understand issues and put these issues into a broader context (Carragee and Roefs, 2004; Oliver and Johnston, 2000; Williams, 2012). Particularly significant in the context of activism, framing shapes how people evaluate the political action they should take (Carragee and Roefs, 2004; Kruse 2001). Framing analysis means focusing on 'what is being said and how it is said' (Rajagopal, 1991, p. xi, cited in Carragee and Roefs, 2004), including aspects such as interpretation, selection, exclusion, emphasis, placement, visual effects and labelling (Parenti, 1993; Reese, 2001), which were considered in the following analysis.

While much of the research cited above explores framing in the context of media analysis, it can also be used to understand messaging from social movements. Of course, the messaging of activists is sometimes featured in more mainstream media (Boyd-Barrett, 2006) through interviews and coverage of protest actions, so they are able to shape the frames of the media in that way, as well as create their own frames through social media and signage at rallies, with this signage being the particular focus here. Carrie Freeman (2013) points out that social movements construct frames that define problems, suggest solutions and encourage participation. It is important to focus on framing in order to gain a greater understanding of social movements such as the animal advocacy movement (Munro, 2012). William Gamson (2004, pp. 245, 247) explains that 'Today, we recognize contests over meaning as framing contests' and that 'competition among frames within a movement about which one should be promoted and emphasized is one major component of a frame-critical analysis of movements'. The contests between different animal activists regarding who should be portrayed as responsible for animal exploitation and the actions that should be promoted to the public provide a clear example of a contest over meaning, which will be explored below.

Existing research exploring the frames presented by animal activists includes a range of work showing that animal advocacy organizations have tended to frame the problem of our relationship with other animals in an animal welfare-based way, emphasizing the suffering of animals due to cruelty, rather than using more rights-based frames, such as opposing animal slaughter and the commodification of animals as objects in contrast to subjects (Freeman, 2010; Pendergrast, 2015; Williams 2012). Another example is James Jasper and Jane Poulsen's (1995) research, which analysed how the frames presented by animal advocates affect whether new people join the movement or not. Finally, Marie Mika (2006) used focus group research to investigate similar issues, focusing particularly on PETA's controversial campaigns and how they are perceived by people outside of the animal advocacy movement. The following framing analysis will build on this literature by also suggesting ways in which animal activists can change their messaging to account for the structural causes of animal exploitation discussed above.

Environmental activism challenging animal agriculture: Vegan Rising

In September 2019, it was estimated that almost 300,000 people participated in coordinated marches around Australia, as part of the global School Strike for Climate marches (Ilanbey, Grieve and Sakkal, 2019). The biggest rally was in Melbourne, where organizers reported an attendance of 100,000 people, making it one of the largest rallies ever held in the city (Ilanbey, Grieve and Sakkal, 2019). The messaging from animal activists at this rally is explored due to the significance of the rally. The banner focused on in the framing analysis (Figure 5.1) is specifically discussed because it is typical of the individual-level analysis dominant in the animal advocacy movement (Wrenn, 2018) and also due to the prominence of this banner at the rally. The banner read, 'How can we expect change from others when we refuse to change ourselves?' This banner was displayed by the organization Vegan Rising, which is perhaps best known for its role in the Dominion Animal Liberation disruption discussed above.[1] From my participation in this Melbourne march (though not as part of the Vegan Rising presence), I can attest that this banner from Vegan Rising was the biggest, most visible banner from the 'vegan contingent' at the rally. Due to the size of the banner, which dominated all other messaging from those highlighting the environmental impact of animal agriculture, it is likely

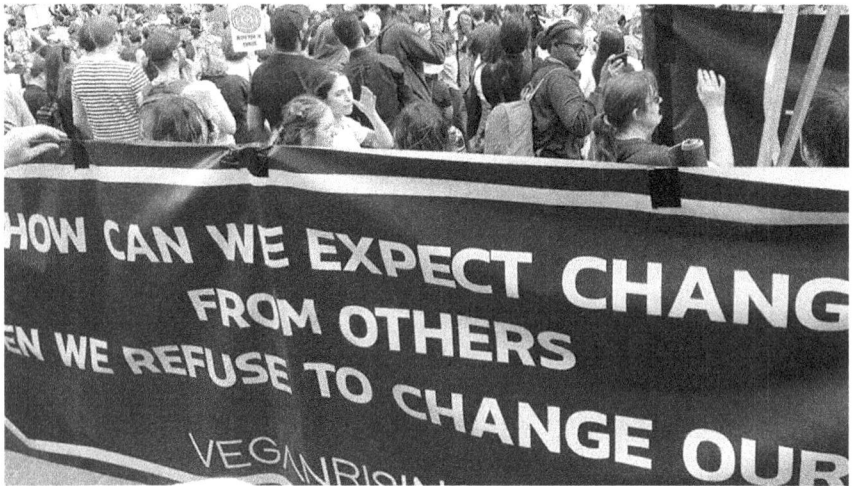

Figure 5.1 Vegan Rising's banner at the School Strike for Climate march in Melbourne.
Source: Vegan Rising (2019).

to be viewed as the most representative of the messaging of these activists overall. This is similar to the way the story on the front page of a newspaper is framed as the most important, due to its heightened visibility (Pendergrast, 2008).

The following framing analysis is not intended to dismiss the validity of switching to a plant-based diet on environmental grounds. In fact, sustainability researchers Seth Wynes and Kimberly Nicholas (2017, pp. 4–5) found that eating a plant-based diet is a 'high-impact' action an individual can take to challenge climate change. This is in contrast to more widely advocated but less-significant individual measures, including 'moderate-impact' actions such as switching from a petrol car to a hybrid, washing clothes in cold water, recycling or hang-drying clothes, as well as 'low-impact' actions such as upgrading light bulbs. This analysis is also not intended to reject individual change more generally – as discussed earlier, both individual and structural responses to environmental and other issues are important.

However, I question whether the messaging in the banner above, which not only highlights individual responsibility but focuses particularly on a hypocrisy narrative, is likely to be useful in convincing non-vegan environmentalists. This is particularly the case because this framing is actually fairly similar to critiques of these climate marches from those who are climate change deniers and/or on the right and are fundamentally *opposing* these actions. Examples include Malcolm Roberts (2019), a climate change denier and politician from the far-right Australian political party One Nation, suggesting that the structural demands of the march were invalid because some participants were using plastic water bottles. There were similar critiques from other right-wing commentators such as Alan Jones (cited in Chung, 2019) and Amanda Vanstone (2019), suggesting the protestors should instead take individual actions to help the environment. Actions recommended included stopping charging their phones and iPads, stopping using air-conditioning and not owning cars, rather than focusing on the contribution of governments and companies to the climate crisis. While all of these individual steps are positive actions that people can take for the environment, the extent to which those in these movements take these *individual* actions really has very little to do with the validity of the *structural* demands from the protestors to stop the Adani coal mine (in Queensland, Australia) and to move towards renewable energy. These were both central demands from the movement, which will be discussed in more detail below.

There are philosophical problems with this hypocrisy narrative expressed in the banner – as discussed earlier, it is easier for some than others to live a vegan lifestyle. There are also more practical issues – highlighting hypocrisy is not necessarily an effective strategy in bringing about behaviour change. Meat-related cognitive dissonance is relevant here; this refers to the discomfort felt by individuals who both consume meat but also recognize the harm caused by meat consumption (Rothgerber and Rosenfeld, 2021). In order to avoid this discomfort, a wide range of research reviewed by psychologists Hank Rothgerber and

Daniel Rosenfeld (2021, p. 5) has found that individuals are much more likely to employ 'motivated justifications, rationalizations, and other strategic attitudes that explain away the troubling nature of eating animals', rather than changing their behaviour to stop eating animals.

I will suggest some ways in which environmental vegan activists can make the case that animal agriculture is a significant environmental issue while avoiding feeding into this hypocrisy narrative. These suggestions are based on the structural focus of sociology and also draw lessons from other social movements that could inform the messaging of animal activists. One example is environmental protest movements generally, and specifically the School Strike 4 Climate Australia (2019) movement, which has the following three demands:

1 No new coal, oil and gas projects, including the Adani mine.
2 100 per cent renewable energy generation and exports by 2030.
3 Fund a just transition and job creation for all fossil-fuel workers and communities.

The focus on fossil fuels but not animal agriculture in these demands has been justified, including by climate organizers who accept that animal agriculture has a significant environmental impact, because they don't want to alienate people who are not vegan (e.g. Coco, 2019). However, people within these movements not only generally consume animal products but also use fossil fuels (to different levels), and some even work within these industries (Peacock, 2019). So why are these demands regarding fossil fuels generally considered non-alienating, while addressing animal agriculture is often viewed as excluding or even hostile towards the many people in environmental movements who currently consume products from animal agriculture? This relates to the *structural* nature of these demands about fossil fuels. Individuals in these movements all consume different levels of fossil fuels, but the demands are not about individual choices. So individuals, regardless of their level of fossil fuel consumption, can unite behind these structural demands.

In contrast, demands concerning addressing the environmental impact of animal agriculture are often framed in a very individual-focused manner, as discussed above. However, I question whether this has to be the case. For example, if animal agriculture was addressed in exactly the same structural way as fossil fuels, I argue that this issue would be more likely to be raised in the School Strike 4 Climate marches and in environmental protest movements generally. Rather than being focused on individual veganism, the demands around animal agriculture could be the following, based on the existing School Strike 4 Climate demands around fossil fuels:

1 No new animal agriculture.
2 100 per cent plant-based agriculture by 2030.
3 Fund a just transition and job creation for all animal agriculture workers and communities.

Such messaging could be used instead of or in conjunction with more individual-focused vegan messaging. As will be discussed in the section 'Reframing Vegan Outreach' below, such promotion of individual change can still highlight the structural cause of the problem, rather than framing the issue purely in terms of individual responsibility. Avoiding the hypocrisy narrative can assist in making environmental vegan activists more a *part* of environmental movements, rather than as an *oppositional* force to these movements. The structural demands highlighted above would be more likely to be accepted by members of environmental movements and would also be more likely to be incorporated into official demands as they fit within the existing framing of these movements.

While animal advocates can certainly learn from other social movements, which will be discussed in more detail below, they should also not unquestionably accept the messaging of these movements. For example, social movement researcher Kyle Matthews (2020a) has looked critically into the messaging of the environmental organization Extinction Rebellion. He questions Extinction Rebellion's argument that a social movement achieving 3.5 per cent of the population of a country engaged in sustained protest guarantees success, a claim the organization bases on research by Chenoweth and Stephan (2012). Matthews (2020a) challenges this claim because Chenoweth and Stephan's research is based on activism against dictatorial regimes, rather than the context of liberal democracies, within which Extinction Rebellion primarily operates. In Australia, environmental movements have carried out mass climate protests, including from Extinction Rebellion and the School Strike 4 Climate marches discussed above. Nevertheless, large numbers of people uniting behind structural demands is no guarantee of success, as can be seen with the Adani coal mine going ahead despite widespread opposition. These issues persist even with a recent change of government, as Australian Prime Minister Anthony Albanese (cited in Morton, 2022) has said his Labor government will approve further new fossil fuel export projects if they 'stack up environmentally'.

Social change is complex, and when it comes to the environmental movement, 'there is no number that can be mobilized for guaranteed success' (Matthews, 2020a, p. 607). Matthews (2020a) argues that Extinction Rebellion's misuse of social movement research has meant that they have pursued a strategy of mass civil disobedience that may be less effective than other tactics, including lobbying elites and direct action against fossil fuels. He advocates for tactical diversity and argues that attempting to create mass civil disobedience should not mean ruling out but also engaging in other strategies at the same time. These are important critiques of the messaging of environmental movements. However, while mass protests should not be fetishized and do not guarantee success, environmental vegan activists engaging in structural messaging based around a mass movement would not mean they are limited to this as their only tactic. In fact, adopting this approach would contribute to the diversity of tactics in the animal advocacy movement, which has relied primarily on individual consumer-based change (Wrenn, 2018). Bringing in more structure-orientated messaging can also allow for more effective messaging that promotes individual change, which will be discussed in more detail below.

The structural messaging around animal agriculture advocated for above would mean reducing the barrier to participation by not mandating particular individual behaviours, which is consistent with the mass participation desired by the organizers of such events. McCarthy and Zald's (2001) 'little effort paradigm' suggests that in terms of maximizing participation in social movements, actions that require the least time and effort are most likely to be pursued by the largest number of people. Attending a rally requires a higher level of commitment than other less-time-consuming actions such as signing a petition or making a donation to an organization (McCarthy and Zald, 2001). However, veganism requires a much higher level of commitment than attending a rally. As discussed throughout this chapter, the difficulty of veganism depends on the specific circumstances of each individual, however, it certainly requires more of an ongoing commitment than simply attending a rally for a few hours, so if this was a requirement to take part in these actions, it would be likely to reduce participation (Pendergrast, 2021). With the structural demands around animal agriculture outlined above, similar to fossil fuels, people within these movements would all consume different levels of animal products (with the less the better for the environment and animals), but many could unite behind these *structural* demands on the importance of moving away from animal agriculture at the government, industry and society levels.

This approach would take advantage of the broad *philosophical* support for veganism, despite the fact that only a small proportion of people live a vegan lifestyle in practice (Bryant, 2019). This *in principle* support for veganism is demonstrated in a wide range of studies, including a survey of people in the United Kingdom by psychologist Christopher Bryant (2019), which found that a majority of meat eaters viewed veganism as ethical and good for the environment. However, these same respondents viewed the *practical* aspects of veganism in a negative way, such as 'ease, convenience, taste, enjoyableness and affordability' (Bryant, 2019, p. 7). Specifically on the environment, a survey of 1.2 million people in 50 countries, the largest ever opinion poll on climate change, by the UN Development Programme (UNDP 2021) found that 30 per cent of people worldwide supported the promotion of plant-based diets to address the climate crisis. On animal concerns and veganism, a 2020 survey by the Sentience Institute found that among the 1,113 participants from the United States, '44.8% are in favor of banning slaughterhouses', despite most of these people eating animals themselves (Ladak and Anthis, 2021). This survey has a much smaller sample size than the climate poll, but these studies taken together show the widespread support for *structural* policies addressing animal agriculture, even if individual consumption does not always totally reflect this. Encouraging individual consumption in line with these attitudes is also worth pursuing; however, such advocacy can also be reframed to highlight the broader structural factors behind animal exploitation.

Reframing vegan outreach: Melbourne Cow Save's Anti-Dairy Campaign

Melbourne Cow Save's anti-dairy campaign involves participants standing on the steps of a busy intersection in the Bourke Street pedestrian mall in Melbourne holding signs with slogans highlighting the plight of 'bobby calves' in the dairy industry. It ran regularly for many years until, like many activist campaigns, it was put on hold during the coronavirus pandemic. The term 'bobby calves' refers to newborn calves who are less than 30 days old and have been separated from their mothers (RSPCA, 2021). The RSPCA (2021) explains that 'For cows to produce milk, they have to give birth to a calf'. The male calves are not useful to the dairy industry because they do not produce milk and in Australia can be legally slaughtered from five days old (RSPCA, 2021). From my own participation in this campaign, I can add that in addition to the signs displayed, this campaign also involves vegan outreach, with information distributed to the public about the problems of dairy and also promoting the consumption of dairy-free alternatives and veganism more broadly. A small group of selected activists distributes this information, as well as engaging in discussions with the public about the issue, with an information stall that has pamphlets for further information. The campaign additionally sometimes involves the distribution of complementary dairy-free alternatives such as dairy-free milks and chocolates. One really positive aspect of this campaign is the accessibility in terms of allowing easy participation – those taking part can simply hold a sign for as long as they'd like to. This is inclusive in terms of reducing barriers to participation for people with disabilities (Priestley et al., 2016), as well as those with other health issues or a lack of time or confidence when it comes to answering questions about dairy and veganism.

This campaign involves very individually focused messaging, with placards highlighting the plight of bobby calves featuring slogans pictured below such as 'He died for your yoghurt', 'He was killed because you wanted cheese' and 'He died for your whipped cream' (Melbourne Cow Save, 2018) and 'Whose baby should die for your dairy? Hers? Hers? Hers?', with pictures of mother cows (Figure 5.2). In terms of framing, consistent with the above example from Vegan Rising, it portrays individuals as *solely* responsible for animal exploitation – these animals are portrayed as being killed due to the public's demand for animal products, typical of the supply and demand model widely articulated in the animal advocacy movement (Wrenn, 2018). Likewise, typical of the messaging in the animal advocacy movement discussed above, individual solutions are offered to deal with the problem, with individuals encouraged to refrain from consuming dairy and other animal products. In rethinking such messaging, lessons can be learnt from other social movements.

Framing in other social movements

The animal advocacy movement is by no means the only movement that promotes individual change. This messaging is consistent with the broader movement of ethical consumption, which is all about consuming products that

Figure 5.2 Photos of Melbourne Cow Save's anti-dairy campaign. The image on the left is from Melbourne Cow Save (2018) and the one on the right was taken by the author.

havea less impact on not only other animals but also on humans and the environment (Harper and Makatouni, 2002; Schröder and McEachern, 2004). The increasing focus on consumer solutions to social problems, where society is shaped by 'the consumer's economic vote', can be linked to the growing dominance of neoliberalism (free market capitalism) (Wrenn, 2011, pp. 17–19). There are many barriers to ethical consumption serving as a means of political activism, including increased cost in many cases (María, 2006; Wrenn, 2011). Despite these barriers, individual actions can at least be *part* of bringing about broader social change. As noted above, the purpose of this chapter is not to totally reject individual change, but rather to question the almost exclusive reliance on it within the animal advocacy movement, as well as to suggest ways in which messaging promoting individual change can be reframed.

While there are certainly examples of other social movements with similar messaging around individuals as being both the source of the problem and the solution, such as discussions around individual carbon footprints (Gram-Hanssen and Christensen, 2012), there are also examples that promote individual change while pointing to the source of the problem being elsewhere. One example is the environmental campaign to encourage individuals to avoid having an account with a bank that invests in fossil fuels. While this is a consumer boycott, with individual actions portrayed as a solution to the problem, individuals are not framed as the *cause* of the problem. Rather, it is broader, structural forces that are portrayed as responsible for these carbon emissions – in this case, banks investing in fossil fuels. Messaging includes 'Is your bank investing in dirty fossil fuels?', with 'your bank' underlined (Figure 5.3) (Market Forces, 2015).

The Boycott, Divestment and Sanction (BDS) movement that challenges Israel's occupation of Palestine is another example that includes the promotion of individual change while portraying structural forces as being responsible for the harm. This campaign includes a boycott of 'all Israeli and international companies engaged in violations of Palestinian human rights' (BNC, 2021); it has some parallels with veganism in that it is an individual economic boycott attempting to challenge an injustice. However, while the focus is on individual change, it is not individuals consuming products from these companies who are framed as the *cause* of the suffering of the Palestinian people. Rather, it is the companies themselves and the Israeli government who are framed as responsible for the harm; individuals can be a part of the *solution* by joining this campaign. Messaging at rallies reflects this structural blame and the positive role of individuals, with slogans such as 'Boycott Israel BDS' (Hauss, 2017) and 'Divest For Palestine' (Rizvi, 2020). Such messaging is likely to attract greater support than messaging that portrays individuals as the cause of the problem, as it can overcome the defensiveness identified above in the points on meat-related cognitive dissonance (Rothgerber and Rosenfeld, 2021).

Melbourne Cow Save's anti-dairy campaign and animal advocates engaged in vegan outreach generally could incorporate similar messaging to these other social movements when encouraging individual change. The animal advocacy

Figure 5.3 An environmental campaign encouraging people to question who they
        bank with.
Source: Market Forces (2015).

movement should explore ways to acknowledge the structural forces that contribute to animal exploitation, such as companies and governments (Wrenn
2018), in their messaging. In the case of this anti-dairy campaign, rather than
the very individual messaging outlined above, such as 'He was killed because
you wanted cheese', a sociological approach that accounts for the influence of
structures might posit the alternative: 'The *dairy industry* kills cows, boycott
this slaughter industry!' This messaging highlights the problems of animal
slaughter within the dairy industry, portraying this industry, rather than individuals, as creating and being responsible for this problem. More positively,
individuals are simply portrayed as part of the *solution* to the problem, being
encouraged to participate in an economic boycott of the industry, similar to
the anti-fossil fuel and BDS campaigns outlined above. As with environmental
campaigning around animal agriculture, those promoting veganism for animals can incorporate structural messaging into their framing.

Conclusion

This chapter has analysed the framing of Vegan Rising in its environmental
vegan advocacy and Melbourne Cow Save in its animal-focused vegan outreach. These examples were chosen because they are representative of the
individual-focused framing that dominates animal advocacy (Wrenn, 2018).

Despite this messaging, animal industries play a substantial role in creating demand for animal products – for example, through advertising. Governments also often support animal exploitation by making it legal, regulating and legitimizing the process, giving subsidies to animal agriculture and criminalizing activists who interfere with the profits of animal-exploiting industries (Wrenn, 2018). Animal advocates can benefit from incorporating a sociological perspective that accounts for both individual agency *and* broader structures that influence individual choices and constrain agency (Cudworth, 2005). Taking on these ideas would make animal advocates cautious about drawing on a hypocrisy narrative in their messaging, as this narrative is flawed philosophically in terms of overemphasizing individual agency, but also practically, in that pointing out hypocrisy does not usually lead to behavioural change (Rothgerber and Rosenfeld, 2021).

Applying this interplay between structure and agency, which is a central theme in sociology, to messaging around veganism is breaking new ground in the sociological literature, as well as providing insight into common tactics used in the animal advocacy movement. This chapter adds to this literature on sociology and animal advocacy by suggesting ways in which campaigns can incorporate the concepts from this literature and the structure–agency debate within sociology generally. This focus on encouraging more effective vegan messaging makes this chapter a substantial contribution to the developing field of vegan sociology (International Association of Vegan Sociologists, 2021). This contribution is particularly important in light of the urgency of a societal shift away from animal consumption in the context of the zoonotic transmission of COVID-19 and the significant role played by animal-based foods in the climate crisis.

Note

1 The Dominion Animal Liberation action occurred just a few months before this rally, and in addition to the blockades of slaughterhouses discussed above, the action also most famously involved blocking one of Melbourne's busiest intersections for nearly three hours during morning peak hour traffic (Hope, Webb and Bourke, 2019).

References

Animal Justice Party. (2017). AJP policies. https://animaljusticeparty.org/wp-content/uploads/2017/11/AJP-Policies16-1.pdf
Animal Rebellion. (2021a). Animal Rebellion stages overnight McDonald's occupation amid call for plant-based food system. *Media release*. https://animalrebellion.org/animal-rebellion-stages-overnight-mcdonalds-occupation-amid-call-for-plant-based-food-system
Animal Rebellion. (2021b). Climate activists blockade UK's biggest milk factory calling for an end to dairy. *Media release*. https://animalrebellion.org/climate-activists-blockade-uks-biggest-milk-factory-calling-for-an-end-to-dairy
Animal Rebellion. (2021c). Homepage. https://animalrebellion.org

Baran, S. (2017). Visual patriarchy: PETA advertising and the commodification of sexualized bodies. In D.A. Vachoch and S. Mickey (eds), *Women and Nature? Beyond Dualism in Gender, Body, and Environment*. New York: Routledge.

BNC. (2021). What is BDS? Palestinian BDS National Committee. https://bdsmovement.net/what-is-bds

Bogueva, D. and Schmidinger, K. (2021). Normality, naturalness, necessity, and nutritiousness of the new meat alternatives. In Information Resources Management Association (ed.), *Research Anthology on Food Waste Reduction and Alternative Diets for Food and Nutrition Security*. New York: IGI Global.

Boyd-Barrett, O. (2006). Alternative reframing of mainstream media frames. In D.K. Thussu (ed.), *Media on the Move: Global Flow and Contra-Flow*. London: Routledge.

Bruno, M., Thomsen, M., Pulselli, F.M., Patriizi, N., Marini, M. and Caro, D. (2019). The carbon footprint of Danish diets. *Climatic Change*, 156, 489–507.

Bryant, C.J. (2019). We can't keep meating like this: Attitudes towards vegetarian and vegan diets in the United Kingdom. *Sustainability*, 11(23), 1–17.

Carragee, K.M. and Roefs, W. (2004). The neglect of power in recent framing research. *International Communication Association*, 54(2), 214–33.

Chenoweth, E. and Stephan, M.J. (2012). *Why Civil Resistance Works: The Strategic Logic of Nonviolent Conflict*. New York: Columbia University Press.

Cherry, E. (2021). Vegan studies in sociology. In L. Wright (ed.), *The Routledge Handbook of Vegan Studies*. London: Routledge.

Chung, F. (2019). 'Shouldn't they turn off their mobile phones?': Alan Jones rips into climate strike schoolkids. *Courier Mail*. www.couriermail.com.au/technology/shouldnt-they-turn-off-their-mobile-phones-alan-jones-rips-into-climate-strike-schoolkids/news-story/9989550646b8f2d05734472beef030c6

Clark, M., Hill, J. and Tilman, D. (2018). The diet, health, and environment trilemma. *Annual Review of Environment and Resources*, 43, 109–34.

Climate Save Movement. (2021). Plant based treaty. https://plantbasedtreaty.org

Coco, V. (2019). Radiothon! With Violet Coco of Extinction Rebellion. *Freedom of Species*. www.3cr.org.au/freedomofspecies/episode-201906161300/radiothon-violet-coco-extinction-rebellion

Constable, N. (2009). The commodification of intimacy: Marriage, sex, and reproductive labor. *Annual Review of Anthropology*, 38, 49–64.

Cudworth, E. (2005). Complex systems: 'Nature,' 'society' and 'human' domination. In E. Cudworth (ed.), *Developing Ecofeminist Theory: The Complexity of Difference*. London: Palgrave Macmillan, 42–70.

Deckha, M. (2008). Disturbing images: Peta and the feminist ethics of animal advocacy. *Ethics & the Environment*, 13(2), 35–76.

Donaldson, S. and Kymlicka, W. (2015). Farmed animal sanctuaries: The heart of the movement? A socio-political perspective. *Politics and Animals*, 1(1), 50–74.

Freeman, C.P. (2010). Framing animal rights in the 'Go Veg' campaigns of US animal rights organizations. *Society & Animals*, 18(2), 163–82.

Freeman, C.P. (2013). Stepping up to the veggie plate: Framing veganism as living your values. In E. Plec (ed.), *Perspectives on Human–Animal Communication: Internatural Communication*. New York: Routledge.

Gamson, W.A. (2004). Bystanders, public opinion, and the media. In D.A. Snow, S. Soule and H. Kriesi (eds), *The Blackwell Companion to Social Movements*. Malden, MA: Blackwell.

Germov, J. and Hornosty, J. (2009). Imagining health problems as social issues. In J. Germov (ed.), *Second Opinion: An Introduction to Health Sociology*. Oxford: Oxford University Press.

Germov, J. and Poole, M. (2015). The sociological gaze: Linking private lives to public issues. In J. Germov and M. Poole (eds), *Public Sociology: An Introduction to Australian Society*. Sydney: Allen & Unwin.

Glasser, C.L. (2011). Moderates and Radicals Under Repression: The US Animal Rights Movement, 1990–2010. PhD thesis, University of California, Berkeley, CA.

Gram-Hanssen, K. and Christensen, T.H. (2012). Carbon calculators as a tool for a low-carbon everyday life? *Sustainability: Science, Practice and Policy*, 8(2), 19–30.

Harper, A.B. (2010). Race as a 'feeble matter' in veganism: Interrogating whiteness, geopolitical privilege, and consumption philosophy of 'cruelty-free' products. *Journal for Critical Animal Studies*, 8(3), 5–27.

Harper, G.C. and Makatouni, A. (2002). Consumer perception of organic food production and farm animal welfare. *British Food Journal*, 104(3), 287–99.

Hauss, B. (2017). The First Amendment protects the right to boycott Israel. *ACLU*. www.aclu.org/blog/free-speech/first-amendment-protects-right-boycott-israel

Hope, Z., Webb, C. and Bourke, L. (2019, 8 April). Vegan protesters unapologetic for Melbourne's peak-hour chaos, abattoir blockades. *The Age*. www.theage.com.au/national/victoria/vegan-protesters-unapologetic-for-melbourne-s-peak-hour-chaos-abattoir-blockades-20190408-p51by5.html

Howard, H. (2021). The beef with vegans: Managing stigma in Britain's hegemonic meat culture. *Geoverse*. www.brookes.ac.uk/geoverse/original-papers/the-beef-with-vegans--managing-stigma-in-britain-s-hegemonic-meat-culture

Ilanbey, S., Grieve, C. and Sakkal, P. (2019, 20 September). 'This crisis, it affects everyone': Organizers say 100,000 at Melbourne's climate strike. *The Age*. www.theage.com.au/national/victoria/protesters-bring-city-to-a-standstill-for-third-climate-change-protests-20190920-p52tfw.html

International Association of Vegan Sociologists. (2021). Guiding principles for the International Association of Vegan Sociologists. www.vegansociology.com/principles

Jasper, J.M. and Poulsen, J.D. (1995). Recruiting strangers and friends: Moral shocks and social networks in animal rights and anti-nuclear protests. *Social Problems*, 42(4), 493–512.

Kruse, C.R. 2001. The movement and the media: Framing the debate over animal experimentation. *Political Communication*, 18(1), 67–87.

Ladak, A. and Anthis, J.R. (2021). Animals, Food, and Technology (AFT) Survey: 2020 update. *Sentience Institute*. www.sentienceinstitute.org/aft-survey-2020

Laurence, C. (2021). Australian MPs call for plant-based climate solutions ahead of COP26. *Nourish Magazine*. https://nourishmagazine.com.au/earth/australian-mps-call-for-plant-based-climate-solutions-ahead-of-cop26

María, G.A. (2006). Public perception of farm animal welfare in Spain. *Livestock Science*, 103(3), 250–6.

Market Forces. (2015). Big banks need to update their fossil fuel policies. *Friends of the Earth Australia*. www.marketforces.org.au/big-banks-need-to-update-their-fossil-fuel-policies/#

Martin, M. and Brandão, M. (2017). Evaluating the environmental consequences of Swedish food consumption and dietary choices. *Sustainability*, 9(12), 1–21.

Matthews, K.R. (2020a). Social movements and the (mis)use of research: Extinction Rebellion and the 3.5% rule. *Interface*, 12(1), 591–615.

Matthews, T. (2020b, 25 May). Coronavirus has changed our sense of place, so together we must re-imagine our cities. *The Conversation*. https://theconversation.com/coronavirus-has-changed-our-sense-of-place-so-together-we-must-re-imagine-our-cities-137789

McCarthy, J.D. and Zald, M.N. (2001). The enduring vitality of the resource mobilization theory of social movements. In J.H. Turner (ed.), *Handbook of Sociological Theory*. New York: Springer.

Melbourne Cow Save. (2018). Photo of Anti-Dairy Campaign. www.facebook.com/MelbourneCowSave/photos/a.1594488564191697/2018199843714046350 2

Mika, M. (2006). Religion, secular ethics and the case of animal rights mobilization. *Social Forces*, 85(2), 915–41.

Mills, C.W. (1959). *The Sociological Imagination*. New York: Oxford University Press.

Morton, A. (2022, 11 July). Labor faces decisions on approval of up to 27 coal developments including greenfield mines. *The Guardian*. www.theguardian.com/environment/2022/jul/11/labor-faces-decisions-on-approval-of-up-to-27-coal-developments-including-greenfield-mines-analysis-shows

Munro, L. (2012). The animal rights movement in theory and practice: A review of the sociological literature. *Sociology Compass*, 6(2), 166–81.

Nation Rising. (2021a). Homepage. https://nationrising.ca

Nation Rising. (2021b). Open letter: Transition Canada to plant-based food system to prevent future pandemics. https://nationrising.ca/open-letter

Oliver, P. and Johnston, H. (2000). What a good idea! Frames and ideologies in social movement research. *Mobilization*, 5(1), 37–54.

Parenti, M. (1993). *The Politics of New Media* (2nd ed.) New York: St Martin's Press.

Peacock, B. (2019). Two words that convinced a fossil fuel worker to join the strikes. *9 News*. www.9news.com.au/national/australia-climate-strikes-tommy-john-herbert-and-strike-organizer-alexa-stuart-address-just-transition-demand-for-coal-workers-and-communities/3c500b58-5b8a-46f6-8fb8-03bf0936359e

Pendergrast, N. (2008). Australian media hegemony and the internet. Paper presented to *Engaging Place(s)/Engaging Culture(s) conference*, Perth. https://hgs.curtin.edu.au/conference-proceedings/engaging-places-cultures

Pendergrast, N. (2015). Live animal export, humane slaughter and media hegemony. *Animal Studies Journal*, 4(1), 99–125.

Pendergrast, N. (2021). The vegan shift in the Australian animal movement. *International Journal of Sociology and Social Policy*, 41(3/4), 407–23.

Petray, T. and Pendergrast, N. (2018). Challenging power and creating alternatives: Integrationist, antisystemic and non-hegemonic approaches in Australian social movements. *Journal of Sociology*, 54(4), 665–79.

Plimmer, G. (2020). In praise of the office: Let's learn from COVID-19 and make the traditional workplace better. *The Conversation*. https://theconversation.com/in-praise-of-the-office-lets-learn-from-covid-19-and-make-the-traditional-workplace-better-138516

Postone, M. (1993). *Time, Labor, and Social Domination: A Reinterpretation of Marx's Critical Theory*. Cambridge: Cambridge University Press.

Priestley, M., Stickings, M., Loja, E., Grammenos, S., Lawson, A., Waddington, L. and Fridriksdottir B. (2016). The political participation of disabled people in Europe: Rights, accessibility and activism. *Electoral Studies*, 42, 1–9.

Reese, S.D. (2001). Prologue – framing public life: A bridging model for media research. In S.D. Reese, O.H. Gandy and A.E. Grant (eds), *Framing Public Life: Perspectives on Media and Our Understanding of the Social World*. Mahwah, NJ: Lawrence Erlbaum.

Rizvi, H. (2020). If they suppress BDS, no other movement is safe. *New Internationalist*. https://newint.org/features/2020/02/18/if-they-suppress-bds-no-other-movement-safe

Roberts, M. (2019). Shocked!! Kids marching for the environment all using single-use plasticwaterbottles.https://twitter.com/MRobertsQLD/status/1174987041303482368

Rothgerber, H. and Rosenfeld, D.L. (2021). Meat-related cognitive dissonance: The social psychology of eating animals. *Social and Personality Psychology Compass*, 15(5), 1–16.

RSPCA. (2021). What happens to bobby calves? https://kb.rspca.org.au/knowledge-base/what-happens-to-bobby-calves

Scheper-Hughes, N. (1995). The primacy of the ethical: Propositions for a militant anthropology. *Current Anthropology*, 36(3), 409–40.

School Strike 4 Climate Australia. (2019). Here are our demands. https://twitter.com/StrikeClimate/status/1174879792140677120

Schröder, M.J.A. and McEachern, M.G. (2004). Consumer value conflicts surrounding ethical food purchase decisions: A focus on animal welfare. *International Journal of Consumer Studies*, 28(2), 168–77.

Smith, J. and Glidden, B. (2012). Occupy Pittsburgh and the challenges of participatory democracy. *Social Movement Studies*, 11(3–4), 288–94.

Smith, K. (2019). Cow and chicken farmers switch to growing mushrooms. *Live Kindly*, www.livekindly.co/cow-chicken-farmers-switch-growing-mushrooms

Thorpe, A. (2020, 3 May). Reclaiming the streets? We all can have a say in the 'new normal' after coronavirus. *The Conversation*. https://theconversation.com/reclaiming-the-streets-we-all-can-have-a-say-in-the-new-normal-after-coronavirus-137703

Tiwari, R. et al. (2020). COVID-19: Animals, veterinary and zoonotic links. *Veterinary Quarterly*, 40(1), 169–82.

UK Parliament. (2021). Support for the Plant Based Treaty. *Early Day Motions*. https://edm.parliament.uk/early-day-motion/58903/support-for-the-plant-based-treaty

United Nations Development Programme (UNDP). (2021). People's climate vote results. www.undp.org/publications/peoples-climate-vote

Vanstone, A. (2019, 26 September). The Greta Thunberg circus has become a complete farce. *Sydney Morning Herald*. www.smh.com.au/environment/climate-change/the-greta-thunberg-circus-has-become-a-complete-farce-20190926-p52v38.html

Vegan Rising. (2019). Vegan Rising's banner at the School Strike 4 Climate march. www.facebook.com/VegansAreRising/photos/rpp.31871754191178/765330260583235/?type=3&theater

Villanueva, G. (2018). *A Transnational History of the Australian Animal Movement, 1970–2015*. Melbourne: Palgrave Macmillan.

Webber, J. (2021). Vegan meat featured on US menus 1320% more times since before COVID-19. *Plant Based News*. https://plantbasednews.org/news/economics/vegan-meat-us-menu

Williams, C. (2012). *The Framing of Animal Cruelty by Animal Advocacy Organizations*. BA(Hons) thesis, University of Maine.

Wrenn, C.L. (2011). Resisting the globalization of speciesism: Vegan abolitionism as a site for consumer-based social change. *Journal for Critical Animal Studies*, 9(3), 9–27.

Wrenn, C.L. (2018). How to help when it hurts? Think systemic. *Animal Studies Journal*, 7(1), 149–79.

Wynes, S. and Nicholas, K.A. (2017). The climate mitigation gap: Education and government recommendations miss the most effective individual actions. *Environmental Research Letters*, 12(7), 1–9.

Ye, Z.-W. et al. (2020). Zoonotic origins of human coronaviruses. *International Journal of Biological Science*, 16(10), 1686–97.

6 Selling veganism in the age of COVID

Vegan representation in British newspapers in 2020

Corey Lee Wrenn

Introduction

It ain't easy being green, as Kermit the Frog so famously said. Sociological research has uncovered a general derogation of vegans at all levels of society, including the personal (MacInnis and Hodson, 2017), the institutional (Greenebaum, 2016), and the cultural (Cole and Morgan, 2011). This negativity has been identified as a key barrier to vegan transition (Markowski and Roxburgh, 2019), which is a particular nuisance given the litany of inequalities associated with non-vegan consumption ('natural' disasters and zoonotic outbreaks such as COVID-19 included). Given that vegan claims-making directly challenges established power structures and capitalist interests, vegan stigmatization and derision are perhaps predictable. Nonetheless, veganism has managed to capture the popular imagination: vegan options are continually increasing in availability across stores and restaurants, while each year growing numbers of participants register for vegan challenges (such as the UK's Veganuary and the Afro Vegan Society's Veguary). The public seems to be considerably more educated about the treatment of other animals in speciesist industries, as well as about the relationship between speciesism and climate change (Sanchez-Sabate and Sabaté, 2019). COVID-19 could offer an additional window of opportunity by lending weight to the seriousness of veganism's claims and underscoring its potential as a sustainable solution to social injustices related to public health and environmental integrity.

This increased attention is remarkable given that traditional news spaces have historically been antagonistic. Critical animal studies scholars have observed that the media frequently protects the interests of the powerful, particularly as media conglomeration has concentrated ownership among a small number of elites (Almiron, Cole and Freeman, 2016). For this reason, social movements that counter power inequities are often misrepresented or outright ignored in mainstream media (Earl et al., 2004; Hocke, 1999). This power is not absolute, however, and to some extent, media producers must negotiate with their consumers. Having persisted for over a century in Britain, the vegan movement has become a cultural mainstay of interest to audiences regardless of the historical misrepresentation or invisibilization of veganism. How have

DOI: 10.4324/9781003257912-9

mainstream news channels adapted? To address this, I offer an exploratory analysis of mainstream UK newspapers to survey the new normal of vegan ideology in a post-COVID society. I expected that the time that had transpired since previous analyses (Almiron, Cole and Freeman, 2016; Cole and Morgan, 2011), in tandem with the mobilizing moment that the pandemic offered, would result in a substantially different media discourse. I conducted a content analysis of articles mentioning veganism published in 2020, the first full year of COVID-19. In contrast to the more pessimistic findings uncovered by Cole and Morgan's (2011) research conducted in the 2010s, the results of this study uncovered a mediascape that is vegan-curious and generally supportive of plant-based living.

Literature review

The mainstream media are important for activists, given their ability to draw attention to that which is frequently invisible to the public, including issues surrounding speciesism and climate change (Happer and Wellesley, 2019). Similar to other movements, the vegan movement has attempted to utilize mainstream channels to raise awareness of injustice, mobilize activists and other resources, and put pressure on industries and policymakers. Movement scholars have been quite clear, however, that these negotiations are not without serious risk (Sampedro, 1997; Shoemaker and Reese, 1996). News coverage is notoriously onerous to control and frequently creates difficulties for activists when goals and claims are misrepresented. The vegan movement is no exception. As Cole and Morgan (2011) report, most newspaper coverage of veganism has treated it as at best barren, boring, or a passing fad, and at worst a dangerous and hostile threat. Freeman (2009, 2016) further argues that major newspapers set the agenda in a way that protects speciesist industries by normalizing violence against other animals, prioritizing agribusiness perspectives, discussing other animals in objectifying terms and generally ignoring the non-human animal rights debate. More recent research supports these findings by illustrating how the state and industries generally go unrecognized as responsible for the societal problems associated with speciesist food industries. For instance, when the connection between animal agriculture and climate change *is* made, it is usually individuals who are targeted as responsible for changing their behaviour (Kristiansen, Painter and Shea, 2021). The marketplace, in other words, dominates the media discourse: industries are protected while individuals are scapegoated and pressured to consume as directed.

The market's control over the media is not especially new, but with the deregulation of the media in the 1990s, a small number of companies would come to control media production (Diamond, 1991). With industrial titans like Alphabet (Google), Meta (Facebook), and Apple in control of the bulk of disseminated news, social justice movements will inevitably find themselves marginalized, particularly if they threaten to destabilize the unequal system that benefits the industry. Invisibilizing or disparaging activists through the media is a powerful means of protecting the status quo, but social movements may employ a number of measures to overcome this challenge. For instance, they may stage elaborate or particularly disruptive protests to force coverage. They might also recruit public relations experts to better control the media's framing of their campaigns. It is also standard practice for movements to utilize their own media channels to fully control their message, although their inability to reach comparably broad audiences could undermine their utility (and credibility) (Foust and Hoyt, 2018).

Intentional engagement with consumerism is another movement strategy of negotiation in an era of media conglomeration. As Chasin (2000) observes of the gay and lesbian movement of the late twentieth century, framing sexuality as a lifestyle congruent with marketplace behaviour helped to integrate the LGBTIQ+ community through the magic of spending power. Alcohol and car advertisements began to target gays and lesbians, pride parades became spending extravaganzas and marriage equality campaigning hastened the arrival of the gay wedding industry. The gritty politics of homelessness, hate crimes and workplace discrimination that also motivated gay rights activism were effectively marginalized. Many other social movements of the twentieth and early twenty-first centuries have also negotiated with the marketplace to achieve success. Few movements seem immune to this. Also consider the feminist movement and its pandering to the 'you're worth it' self-care industry (Zeisler, 2016) or the environmental movement's 'sustainable' shopping approach. Even the natural foods movement abandoned its push for widely available, nutritious and whole foods for the more profitable shelf-stable vitamin and supplement business (Miller, 2017). Gopaldas (2014) argues that 'marketplace sentiments' (including excitement over new products or outrage over problematic products) can be instrumental in shifting culture (and ultimately even markets themselves). But this relationship with the market is a reciprocal one: consumer emotions are frequently keyed by industries to push products branded as especially ethical. Not long after a movement establishes a social justice campaign to challenge problematic consumption, companies will adapt to align their products with the movement's framework. Gradually, the boundary between movement and marketplace will fade.

Are vegans consuming to liberate other animals or simply consuming in response to corporate advertising? This question will not be answerable in this study, but it is worth considering how corporate operations work to undermine anti-speciesist efforts. It is one thing to blatantly disparage a movement in order to undermine it; it is another altogether more insidious tactic to disrupt a movement's potency by repackaging its politics for sale. In such cases, the power structures remain intact, profits continue to funnel to industry elites and non-human animal liberation is likely to remain a pipedream.

Methodology

This study examines veganism in the British mediascape as it navigates a moment of unique popularity, movement marketization and waves of coronavirus. In a study examining Google Analytics, the United Kingdom was found to be the most popular country for veganism in 2020 (Chef's Pencil, 2021). Although it would certainly be interesting to cover vegan representation in countries where veganism is less popular, this study seeks to understand the depiction of veganism in a region where it has been more or less successful and thus posed a challenge to mainstream media, which must balance the interests of the public with the interests of its industry funders. For that matter, veganism has a more robust history in the United Kingdom than elsewhere in the world, with the Vegan Society having formed there in 1944.

Although a number of measures might have been taken to understand mainstream interpretations of veganism, an analysis of newspapers presents certain advantages: it allows for a wide survey across many regions of the country, it is (at a time of persistent travel and social contact restrictions) a considerably more convenient methodology and it allows me to replicate the Cole and Morgan (2011) study, one of the first (and only) existing studies to test vegan media representation. Cole and Morgan's analysis (which relied on a sample of articles published in 2007) finds that mentions of veganism in British newspapers were few and predominantly negative. In the 13 years since, I expected that this representation would have changed, given the persistence of vegan campaigning, the popularity of Veganuary, and the omnipresent pandemic.

I opted to extract a sample across 2020, the first full year of the pandemic and one of the first years to report robust numbers of Veganuary registrants and vegan product releases. A keyword search in LexisNexis for 'vegan' in UK newspapers published between 1 January and 31 December 2020 yielded 41,175 results (excluding 14 in languages other than English). Some leading news sources from the Cole and Morgan study (conducted in 2007) did not surface (*Daily Express*, *News of the World*, and the *Sunday Telegraph*), while Sunday editions that ranked separately from the daily versions in 2007 are now collapsed (The *Mirror* and the *Daily Express*). Only 19 newspapers covered veganism in the Cole and Morgan study, but dozens covered veganism in the search I conducted 13 years later in 2020. Due to the dramatic increase in vegan coverage, I was unable to code every search result, as was possible in the Cole and Morgan study; instead, I coded the first 35 results for each month of 2020 to achieve an approximate sample of 1 per cent (n = 420).

I based my coding scheme on the approach devised by Cole and Morgan (2011), beginning with three basic categories of positive, neutral and negative. This original study also had several subcodes for negative coverage, given that most of its results were negative. Articles that were deemed negative treated veganism as ridiculous, ascetic, difficult, a fad, dangerous, overly sensitive, or hostile. These codes were reused in my analysis, but I found much more diversity in coverage than Cole and Morgan, such that my coding frame required additional subcoding for positive results as well. Articles in my study that were coded as positive related to product spotlights, weight loss, climate-friendliness, healthfulness, deliciousness, ethics, and how to transition to veganism.

Although the sample was coded only by myself, a colleague was enlisted to test my coding reliability across 10 per cent of the sample. This resulted in the need to clarify some elements of the coding frame. The issues were primarily twofold. First, it was unclear whether an article could be coded as *both* positive and negative; ultimately, I decided to pick only one primary code by determining the predominant tone of the articles. Several articles that would otherwise be positive started off with a negative statement about veganism being bland or hard to cater for, as though there was a need to defend the featured vegan product or recipe from vegan stereotypes. These would be coded as positive only if the positive significantly outweighed the negative. If the article was not clearly leaning either way and might be considered 'balanced', it could be coded as neutral. Second, it was not always clear whether an article would be coded as positive if it focused on products or services for sale. Many such articles in the Cole and Stewart study were coded as neutral, but I opted to code them as positive when there was a clearly positive spin to the product. The final dataset from my sample is available on Figshare.com.[1]

Results

Coverage by newspaper

The most impressive result from this study was the sheer magnitude of vegan coverage. Cole and Morgan's 2007 analysis found only 397 results. My 2020 analysis found 41,175, a dramatic *103.7 times* that of the earlier study. It was quickly clear that, in just the span of a few years, the vegan discourse in mainstream British media had expanded considerably to the point of normalization. All varieties of British newspapers surfaced in the sample, but *Chronicle Live* (3.1 per cent), *MailOnline* (13.3 per cent), the *Mirror* (11.9 per cent), the *Guardian* (2.6 per cent), the *Independent* (8.6 per cent), the *Sun* (5.5 per cent), and the *Times* (3.3 per cent) were the most prominent with at least 10 articles each in the sample (Table 6.1). Several other newspapers, such as the *Scottish Daily Mail* and the *East Anglian Daily Times*, had moderate representation of between six and 10 articles. Dozens of other smaller, local newspapers surfaced with five or less articles represented in each, collectively representing 45 per cent of the sample.

Of the most represented newspapers, right-leaning newspapers (the *MailOnline*, the *Scottish Daily Mail*, the *Sun*, and the *Times*) were about as equally likely to discuss veganism positively as they were negatively, although most of the positive coverage was about new, exciting, or delicious vegan products and services. Of left-leaning journals, the *Guardian* was much less likely to feature veganism, and this coverage was overwhelmingly neutral (usually mentioning vegan products for sale matter-of-factly without hyping up the tastiness, healthiness, launch success, or growth of

Table 6.1 Frequency of discourses of veganism by newspaper

Newspaper	Positive		Neutral		Negative		Total	
	N	%	N	%	N	%	N	%
Chronicle Live	5	1.1	7	1.6	1	0.2	13	3.1
East Anglian Daily Times	7	1.6	1	0.2	0	0.0	8	1.9
Macclesfield Express	4	0.9	4	0.9	1	0.2	9	2.1
MailOnline	25	5.9	4	0.9	27	6.4	**56**	**13.3**
Mirror	18	4.3	16	3.8	16	3.8	**50**	**11.9**
Scottish Daily Mail	4	0.9	0	0.0	5	1.2	**9**	**2.1**
Standard	4	0.9	1	0.9	2	0.4	**7**	**1.0**
Guardian	3	0.7	8	1.9	0	0.0	**11**	**2.6**
The Independent	23	5.5	4	0.9	9	2.1	**36**	**8.6**
The Sun	8	1.9	6	1.4	9	2.1	**23**	**5.5**
The Times	7	1.6	1	1.6	6	1.4	**14**	**3.3**
Other	108	25.7	49	11.7	27	6.4	**184**	**44.0**
Total	**216**	**51**	**101**	**25**	**103**	**24**	**420**	**100.0**

veganism). *The Independent* offered the most positive vegan coverage (64 per cent of the 36 articles), but half of these (48 per cent) related to vegan products and services.

Positive and negative portrayals

Across all newspapers, 63 per cent (136 articles) of the positively coded articles were of this kind (28 articles alone discussed Gregg's product line).[2] Of the remaining positive articles, 27 (6.4 per cent of the sample) were predominantly related to ethics (primarily human justice issues), 18 (4.3 per cent of the sample) touted the health benefits of veganism, 16 focused on its deliciousness, eight were concerned with climate protection, eight aimed to assist readers with transitioning to veganism, and the final three discussed veganism's usefulness for losing weight.

Of those articles focused on ethics, 10 related to the Jordi Casamitjana trial whereby veganism had been declared a protected philosophy. Four articles were about food security and food banks for vegans, two were about the importance of offering options for vegans and the remaining 11 (2.6 per cent of the entire sample) discussed veganism as an ethical duty to other animals. However, almost all of these 11 were couched in larger discussions about vegan activists. For instance, a few articles spotlighting the 'Most Beautiful Vegan Over 50' mentioned the contestants' anti-speciesist motivations for going vegan, but ultimately the articles were interested in the activists themselves, not necessarily the non-human animals they represented. By way of another example, a

MailOnline article titled, 'Fed-up Farmer Clashes with a Vegan Protester Dressed Up as a "Violated" Cow to Stop Supermarket Shoppers from Buying Milk' discusses the vegan argument about dairy consumption, but the article overall focuses on the vegan, not the cows (Mourad, 2020). Only one article out of the 420 sampled in this study spotlighted non-human animal rights purely for the sake of the animals. This was a *MailOnline* article titled: 'Poor Lamb! Heartbreaking Footage of a Baby Sheep Shivering in Paddock after the Herd was Sheared is Posted' (Lackey, 2020).

Negative representations of veganism comprised a quarter of the sample, primarily in the *MailOnline*, the *Mirror*, and smaller newspapers. Thirty-three of these articles warned that veganism was dangerous in some way (primarily to one's health or to that of companion animals). Twenty-four of them emphasized the difficulty or impossibility of sustaining a vegan lifestyle, 15 ridiculed veganism, 12 pitted vegans as hostile, 10 described vegans as overly sensitive, five emphasized the asceticism of veganism, and four presented veganism as a fad. These negative subcodes are the same as those devised by Cole and Morgan, but they were comparatively much less populated in my analysis.

Special events

A number of holidays and current events impacted the portrayal of veganism. The aforementioned Casamitjana trial, for instance, surfaced quite a bit, particularly in the early part of 2020 as the trial came to an end. Most of these trial-related articles were positive (one was neutral). The other major 'political' event to grab the headlines involved a self-described vegan who consumed animal products on an episode of *I'm a Celebrity, Get Me Out of Here*. Eight articles on this topic surfaced, all but one of which were negative (usually emphasizing the hostility of vegan audiences).

Christmas solicited 12 articles mentioning veganism (seven of which were positive, three negative, and two neutral). Easter surfaced five times with two positive articles and three neutral. Earth Day, Pancake Day, National Barbeque Week, New Year's Day, Ramadan, Sausage Roll Day, Thanksgiving, and Valentine's Day also surfaced. Most of these articles are related to menu items and products available to celebrate the occasion. Special events manufactured by the movement surfaced as well. The most popular was Veganuary, which surfaced in 39 articles (15 of which were predominantly positive, 22 neutral, and two negative), while World Vegan Day pulled in seven articles (all positive). PETA's 'Most Beautiful Vegan over 50' surfaced four times, all positively.

As expected, COVID-19 surfaced somewhat regularly. Twenty-two articles mentioned coronavirus or the associated lockdowns as a motivation for going vegan. Only two of these were coded as negative, 20 were positive, and seven were neutral product mentions. Forty-three mentioned COVID-19 as a reason for changing shopping patterns in favour of veganism. Fourteen were coded as neutral and were mostly product related, 29 were coded as positive, and 15 of these were also product related.

Discussion

The new vegan marketplace

Although half of the sample (51 per cent) was coded as positive, this finding is somewhat tempered by its context. Nearly two-thirds (62 per cent) of the positively coded articles covered product launches. In other words, one-third of the entire sample (136 articles) related to positively framed product launches that usually prioritized describing how tasty, exciting, or fast-selling the products were. Most of the neutrally coded articles (one-quarter of the articles in this study were coded as neutral) *also* related to product launches or vegan options. Likewise, quite a few product launches mentioned the lockdown as an opportunity to create new products or test new services such as takeaway menus and grocery boxes.

The data would indicate that a smorgasbord of delights awaits the vegan consumer, but this emphasis on consumption suggests that veganism is gaining legitimacy through the marketplace rather than political resonance. Considerable articles were designed to facilitate veganism for readers by way of their wallets in featuring recipes with luxurious ingredients, announcing vegan festivals, and highlighting restaurants and delivery services catering to vegans. Several positive articles also emphasized the marketplace opportunities for entrepreneurs and investors, reporting on and projecting market growth. Although some of these articles spoke about the reasons why restaurateurs, developers, or consumers would support veganism, usually the ethical mentions were brief and subsumed within the larger narrative about the marketplace or market goods. A typical example is found in a piece with *The Telegraph* about a commercial chef. This feature in the food and drink section was over 1,200 words long, but the only reference to vegan philosophy was confined to a short sentence and quote: 'But when Omari turned eight he decided to go vegan, after watching a Peta video on animal welfare. "It made me sad how the animals were treated," he recalls. "I decided I couldn't eat meat after that"' (Lawrence, 2020). What most vegan activists would consider the central element of veganism is relegated to a sound bite subsumed within a sales pitch.

Working within capitalism to disrupt capitalism has not come without criticism. This depoliticization of veganism through the marketplace has been a particular bane of many activists (Figure 6.1). In the early years of veganism, convenience items and complex analogues were hard to come by, and this was understood to act as a major barrier to vegan outreach. The vegan marketplace today is vast and new products enter British stores in record numbers every year. However, the marketplace is far from a space of equality: it has always

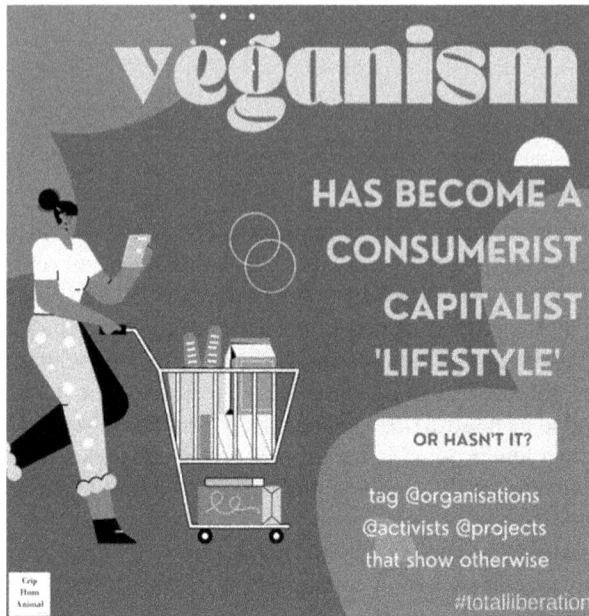

Figure 6.1 Critique of marketplace veganism, Geertrui Cazaux (2021).

favoured society's most privileged, and society's most privileged have worked to control markets to reproduce their power (Chasin, 2000). Freedom, in other words, cannot be for sale. Veganism has shifted from abstention to acquisition; it remains to be seen whether the increasing availability of speciality vegan products can tackle the systemic inequities that maintain human and non-human oppression (Wrenn 2011, 2019a; Wright 2015). The encroachment of 'plant-based' labelling, which is even further divorced from anti-speciesism, only serves to compound this trend. Although this analysis did not include 'plant-based' in the search, I would expect similar patterns of apolitical, consumerist representation.

In any case, vegan campaigners continue to craft marketing events that are sure to grab headlines and encourage spending in an effort to normalize veganism. World Vegan Day, for example, was created by the Vegan Society in the 1990s to promote veganism and its new product labelling scheme (a major source of revenue for the charity) (Wrenn, 2019b) and, based on its appearance in the October and November news coverage surveyed here, it seems to have been successful. Veganuary in January, however, was considerably more popular with the media. Many of the articles discussing Veganuary heralded the coming of new vegan products for sale, as stores and restaurants across the United Kingdom, from Burger King to Superdrug, scrambled to push their new vegan product line in time for New Year's resolutions. Veganuary (itself a registered charity) caters to this sales rush as well, regularly featuring brand sponsors in its newsletter and other outreach material.

Vegan ethics and climate resilience

Whether or not Veganuary, World Vegan Day, or similar events (which rely heavily on branding and corporate sponsorship) are successful is, quite frankly, contingent upon the movement's definition of success. Again, few of the positive articles spoke to animal ethics, but instead prioritized human justice. Typically, these are related to the rise of vegan hate crimes, the availability of vegan food in prisons, the establishment of a vegan food bank in the city of Bath, and the lockdown of food delivery services. Veganism's ability to alleviate climate change (another rather human-centric topic) also appeared more often than animal ethics. Otherwise, there was very little philosophical discussion of veganism, lending credit to concerns that veganism has been coopted for market purposes. This ethical dearth likely relates to the media's alliance with the marketplace and the corporate ownership of news channels; corporations that thrive in a system of inequality would understandably find radical liberatory politics problematic and worth sidelining. On one hand, this decentring of other animals could be seen as a necessary frame adaptation. On the other, it is clearly a major deviation from the original anti-speciesist and liberatory ethic that historically has characterized veganism (Wrenn, 2019a, 2019b).

The COVID-19 factor, likewise, did little to alleviate the invisibilization of non-human animals. Attwood and Hajat (2020) suggest that COVID-19 has disrupted humanity's complacent relationship with 'meat' as it highlighted the dangers of speciesist agriculture and disrupted foodways. A new consciousness was forming, they commented, and this would surely lead to greater acceptance of plant-based approaches. Although speciesism was at the root of the pandemic, and veganism is well situated to alleviate the consequences of COVID-19 and prevent future outbreaks, surprisingly little discussion of this correlation surfaced in the sample. One notable exception was a piece in the *Independent* titled, 'Go Vegan or Risk Further Pandemics, Experts Warn' (Giordano, 2020, p. 2). This piece examines an open letter organized by scientists and Viva! activists,[3] emphasizing the various links between veganism and pandemic prevention. *The Independent* even aligned Viva's initiative with its own pre-existing anti-wildlife trade campaign. It later featured an additional piece written by Viva! Director Juliet Gellatley (2020), which concluded, 'We've known the risks for almost two centuries. Too many lives have been lost. The solution is at our fingertips: it's time to go vegan now'. A spin on this story in the *Mirror* was unsympathetic. Titled 'Man Sparks Outrage by Claiming Coronavirus "Wouldn't Exist If World Was Vegan"', the story primarily consists of vitriolic comments from unqualified readers who dismiss the relationship between human supremacy and COVID-19 (Dresch, 2020).

Persistent veganphobia

The ambivalence about the veracity of veganism's claim to optimal public health was part of a larger vein of media dismissiveness or distrust. Although only a quarter of the sample was disparaging of veganism, its persistence requires attention. Many articles coded as negative seemed intent on counter-framing vegan claims as misleading. For instance, some articles worked to uncover the ingredients and nutritional profiles of vegan substitutes as less healthy and more processed. Indeed, the marketization of veganism seems to have conflated high-calorie convenience foods with 'vegan' food, overlooking the more basic (and healthful) staples of the vegan diet: fruit, vegetables, fungi, nuts, beans, and pulses. Others emphasized the environmental unsustainability of some ingredients associated with veganism, such as avocado, soy, and palm oil.

Of course, veganism has long been criticized by pundits as unhealthy for humans and incongruent with supposedly more sustainable animal-based food systems (how would we have enough plants to eat if everyone went vegan?), but newspapers today grapple with the significantly increased popularity of veganism. Newspapers must now be more accommodating in their attack. As this sample demonstrates, authors hoping to undermine veganism frequently begin an article by acknowledging that veganism is widely understood to be a healthy, earth-friendly diet (thus validating claims that were, in previous years, dismissed outright), before finishing it with refutations. Their new tactic of counter-mobilization is to hype up worst-case scenarios using extreme stereotypes of vegans surviving on infinite slices of environmentally taxing avocado toast and high-sodium Gregg's vegan sausage rolls.

Some of this coverage relied on false claims and lacked scientific substantiation, a tactic that may have currency in the post-Trump, post-truth media age. One 2 April feature in the *Bath Chronicle* titled 'An All-Vegan World Would Be Disastrous' jeers 'Where's the proof that vegan meals are more healthy?' It also warns that 'the space required to sustain a global all-vegan diet would mean the destruction of billions of acres of land and forests, causing disaster for our fragile climate balance, increased air-pollution and the threat of the dustbowls of the American Mid-West in the 1930s in which droughts could (and sometimes did) destroy entire harvests' (Box, 2020, p. 18). A piece for the *Daily Star* went so far as to blame a vegan protestor who had been plastering Welsh farms with vegan stickers for spreading coronavirus (Torre, 2020: 20). Quite a few articles also emphasized the unhealthfulness of feeding vegan diets to companion animals. The considerable frequency of this topic again suggests an act of sabotage to the vegan argument by framing veganism as just the opposite of what it claims to be. Rather than healthy, veganism is unhealthy; rather than good for the environment, it is deleterious for the environment; rather than helping animals, it hurts them.

Most of the articles coded as negative pitted veganism as dangerous in some way, either as problematic for human health or for threatening public safety. A few were coded as negative not because they disparaged veganism *per se*, but because they discussed a negative experience associated with vegan living. For instance, there were quite a few articles related to the accidental serving of animal flesh to vegans. Although these were not necessarily 'veganphobic', and actually validated vegan concerns about consuming animal products, they contributed to the overall negative light cast on veganism and underscored the 'otherness' of vegans (and one could imagine non-vegan audiences finding these stories humorous). These were coded as consistent with Cole and Morgan's (2011) originally identified theme of asceticism (if the articles made vegans appear concerned with purity and abstinence), overly sensitive (if it was emphasized that the gaffe was not so big a deal as to warrant complaint), difficult (if the articles emphasized the difficulty of accommodating vegans), or hostile (if the vegan consumer or vegan community's response was highlighted as particularly aggressive). One was actually coded as dangerous as the article framed it as a restaurant inspection issue. Ultimately, in light of the predominance of product- and consumption-themed articles, these articles could also be interpreted as aligned with marketplace veganism since they emphasized customer service failings or faulty products.

Are these obnoxious vegan customers here to stay? Veganism as a fad did not significantly surface in this analysis as it did in the Cole and Morgan (2011) study, at least as a means of demeaning veganism or emphasizing its temporariness. However, it *was* quite common for articles to emphasize the suddenness of vegan popularity with terms such as 'growth' and 'trend'. Although this might be seen as a form of faddism, I contend that these articles are more accurately interpreted as positive. Indeed, the trendiness of veganism was commonly noted to introduce new products or services.

Another potentially negative trend surfaced in the coverage of events with shock value, such as extreme comments made by reality show celebrity vegans or, as mentioned above, the appearance of bizarre animal parts showing up in produce, restaurant orders, or other vegan goods. There is a need for some nuance to this point, however. Some articles had shocking headlines that appeared to ridicule or mock vegans (a particularly favourite tactic of the *MailOnline*), but the article would then go on to provide considerable space to the vegan argument. This suggests that shocking headlines and content are probably intended as clickbait and may or may not actually care to disparage veganism. In other words, even the articles not pushing products were still oriented towards the marketplace, hoping to drive revenue through page clicks, shares, and subscriptions.

Conclusion

More than two decades into the twenty-first century, where zoonotic pandemics have become a part of life, and a century since activists began to formally organize a vegan movement, is British society at a tipping point for vegan acceptance? This study examined leading newspapers over the course of a year to ascertain the nature of contemporary vegan representation. The findings support the view that veganism is predominantly presented in a positive light, especially with regard to goods to buy, restaurants to visit, and festivals to patronize. Today's veganism is a more or less normal contender in the marketplace, at least as presented by British newspapers.

This monetized lifestyle veganism was predominantly detached from the pandemic. Although the sample used in this study encapsulated the COVID-19 crisis, much of the vegan coverage appears to have been following the momentum of the *pre-existing* normalization of veganism and the substantial upward trend in vegan product development and availability. Some articles did mention COVID-19, but these largely related to the space the pandemic created for testing new products or launching new services. A few articles mentioned COVID-19 as a motivator for trying veganism or at least consuming more vegan food, and this is consistent with the heightened public attention to health, homesteading, and hobbying associated with the pandemic. However, the importance of veganism for preventing pandemics (and strengthening resistance to disease) could have been emphasized but was not. Veganism as a solution to climate change actually surfaced more than veganism as a solution to pandemics. The somewhat ambiguous concept of climate change offers a considerably more generic frame than the pandemic and, notably, climate change is also consistent with 'green' capitalism's sustainable growth (consumption-centric) model. Perhaps, COVID-19 represents a missed opportunity for campaigners, but it is more likely that veganism understood as a site of resistance to zoonotic disease is considerably more political and thus harder to monetize – at least for the time being.

Ultimately, the focus on vegan products and dietary practices has created a depoliticized image of veganism. Critical discussions of non-human animal rights and speciesism were noticeably absent in the sample, while the capitalist encroachment on veganism was omnipresent. Approximately half of the articles in the sample related to vegan products or services. It would be difficult to imagine half of all mentions of feminism, environmentalism, or civil rights activism in mainstream newspapers as product plugs, but this is business as usual for veganism. Many of the articles that advised readers on making the transition to veganism were actually sales pitches that were strategically annotated with advertisements and product placements. The 'veganphobia' that characterized the early years of the twenty-first century seems to have largely vanished and the antidote was commodification in the marketplace. Recent research, incidentally, has noted that anti-vegan stigma (promulgated by the 'veganphobic' mediascape outlined by Cole and Morgan (2011)) may not actually be so powerful a deterrent. Rosenfeld and Tomiyama (2020), for instance, find that, although concerns with stigma mattered somewhat, respondents were more worried by their preconceptions about the taste and nutrition of plant-based foods. Perhaps, then, the media's emphasis on healthful and delicious new products will be complementary to the movement's prioritization of vegan transition.

If the vegan movement were to align with the anthropocentrism and market politics of British news, the subsequent compromise to ethical claims-making would undoubtedly be cause for concern. Social movements across history have conceded that incorporation into the marketplace is key to achieving platform and legitimization, but this strategy comes at a cost. This sample makes it clear that the radical anti-speciesist ethic of veganism has been completely divorced from the pleasure-seeking, profit-focused consumerism of mainstream newsworthy veganism. Whether this approach can seriously undermine humanity's oppressive relationship with other animals remains to be seen.

Acknowledgements

I would like to thank Casey Riordan for assistance with sourcing literature and Lynda Korimboccus for her assistance in designing and testing the coding frame.

Notes

1 See https://figshare.com/articles/dataset/Veganism_in_the_Age_of_COVID/17275139.
2 Gregg's is a highly popular British fast food chain specializing in the highly traditional sausage roll. It rolled out a vegan version in 2019 with resounding success, encouraging a national conversation and inspiring a variety of other fast food chains to follow suit.
3 Viva! is a mid-sized British vegan charity.

References

Almiron, N., Cole, M. and Freeman, C. (2016). *Critical Animal and Media Studies*. London: Routledge.
Attwood, S. and Hajat, C. (2020). How will the COVID-19 pandemic shape the future of meat consumption? *Public Health Nutrition*, 23(17), 3116–3120.
Box, H. (2020, 2 April). An all-vegan world would be disastrous. *Bath Chronicle*.
Cazaux, Geertrui (2021). About animal resistance and human saviourism. *International Animal Rights Conference*, Luxembourg, Sept 2–5.
Chasin, A. (2000). *Selling Out: The Gay and Lesbian Movement Goes to Market*. New York: Palgrave Macmillan.
Chef's Pencil. (2021). Most popular countries and cities for vegans in 2020. www.chefspencil.com/most-popular-countries-and-cities-for-vegans-in-2020-jan-2021-update
Cole, M. and Morgan, K. (2011). Veganphobia: Derogatory discourses of veganism and the reproduction of speciesism in UK national newspapers. *The British Journal of Sociology*, 62(1), 134–153.
Diamond, E. (1991). *The Media Show: The Changing Face of the News, 1985–1990*. Cambridge, MA: MIT Press.
Dresch, M. (2020, 17 March). 'Man Sparks Outrage by Claiming Coronavirus 'Wouldn't Exist if World was Vegan'. *Mirror*.
Earl, J., Martin, A., McCarthy, J. and Soule, S. (2004). The use of newspaper data in the study of collective action. *Annual Review of Sociology*, 30, 65–80.
Foust, C. and Hoyt, K. (2018). Social movement 2.0: Integrating and assessing scholarship on social media and movement. *Review of Communication*, 18(1), 37–55.
Freeman, C.P. (2009). This little piggy went to press: The American news media's construction of animals in agriculture. *The Communication Review*, 12(1), 78–103.
Freeman, C.P. (2016). This little piggy went to press. In N. Almiron, M. Cole and C.P. Freeman (eds), *Critical Animal and Media Studies*. New York: Routledge.
Gellatley, J. (2020, 30 April). If more of us were vegan, there would be less chance of a pandemic in the future. *The Independent*.

Giordano, C. (2020, 30 April). Go vegan or risk further pandemics, experts warn. *The Independent*.

Gopaldas, A. (2014). Marketplace sentiments. *Journal of Consumer Research*, 41, 995–1014.

Greenebaum, J. (2016). Questioning the concept of vegan privilege: A commentary. *Humanity & Society*, 4(1), 355–372.

Happer, C. and Wellesley, L. (2019). Meat consumption, behaviour and the media environment: A focus group analysis across four countries. *Food Security*, 11, 123–139.

Hocke, P. (1999). Determining the selection bias in local and national newspaper reports on protest. In D. Rucht, R. Koopmans and F. Niedhardt (eds), *Acts of Dissent: New Developments in the Study of Protest*. Lanham, MD: Rowman & Littlefield.

Kristiansen, S., Painter, J. and Shea, M. (2021). Animal agriculture and climate change in the US and UK elite media. *Environmental Communication*, 15(2), 153–172.

Lackey, B. (2020, 7 July). Poor Lamb! Heartbreaking footage of a baby sheep shivering in paddock after the herd was sheared is posted online as vegans call for the wool industry to be axed. *MailOnline*.

Lawrence, M. (2020, 9 March). Meet the 11-year-old chef who founded his own vegan food venture. *The Telegraph*.

MacInnis, C. and Hodson, G. (2017). It ain't easy eating greens: Evidence of bias toward vegetarians and vegans from both source and target. *Group Processes & Intergroup Relations*, 20(6), 721–744.

Markowski, K. and Roxburgh, S. (2019). 'If I became a vegan, my family and friends would hate me': Anticipating vegan stigma as a barrier to plant-based diets. *Appetite*, 135, 1–9.

Miller, L. (2017). *Building Nature's Market*. Chicago: University of Chicago Press.

Mourad, L. (2020, 28 July). Fed-up farmer clashes with a vegan protester dressed up as a 'violated cow' to stop supermarket shoppers from buying milk. *MailOnline*.

Rosenfeld, D. and Tomiyama, A. (2020). Taste and health concerns trump anticipated stigma as barriers to vegetarianism. *Appetite*, 144, 104469.

Sampedro, V. (1997). The media politics of social protest. *Mobilization*, 2(2), 185–205.

Sanchez-Sabate, R. and Sabaté, J. (2019). Consumer attitudes towards environmental concerns of meat consumption: A systemic review. *International Journal of Environmental Research and Public Health*, 16(7), 1220.

Shoemaker, P. and Reese, S. (1996). *Mediating the Message: Theories of Influences on Mass Media Content*. White Plains, NY: Longman.

Torre, B. (2020, 21 April). Farmer fury at Peppa pranks: Vegan posters blasted. *Daily Star*.

Wrenn, C. (2011). Resisting the globalization of speciesism: Vegan abolitionism as a site for consumer-based social change. *Journal for Critical Animal Studies*, 9(3), 9–27.

Wrenn, C. (2019a). *Piecemeal Protest: Animal Rights in the Age of Nonprofits*. New York: SUNY Press.

Wrenn, C. (2019b). The Vegan Society and social movement professionalization, 1944–2017. *Food and Foodways*, 27(3), 190–210.

Wright, L. (2015). *The Vegan Studies Project*. Athens, GA: University of Georgia Press.

Zeisler, A. (2016). *We Were Feminists Once*. New York: PublicAffairs.

Part III
Species(ist) relations

7 Food animals as an economic class

Animals as commodities under capitalism

Dinesh Wadiwel

Introduction

One of the contributions of Karl Marx (1986, 1992) to the development of sociology was in the articulation of a 'structural' perspective on class relations within a society. According to this view, the economic structure of a society determines the collective position of individuals and their access to power; that is, their 'class'. In contrast to other treatments of class, such as Max Weber's interrogation of social status, Marx was less interested in cultural markers of social stratification and in income and wealth distribution *per se*; rather, class positions reflected the orientation of individuals to production processes, and this orientation would prove all-important in determining the social power exercised by these classes, and the relations of antagonism that would follow. Anthony Giddens summarizes:

> Marx's emphasis that classes are not income groups is a particular premise, stated in *Capital*, that the distribution of economic goods is not a sphere separate to and independent of production, but is determined by the mode of production ... Classes are constituted by the relationship of groupings of individuals to the ownership of the private property in the means of production. This yields a model of class relations which is basically dichotomous: all class societies are built around a primary line of division between two antagonistic classes, one dominant and the other subordinate. In Marx's usage, class of necessity involves a conflict relation.
>
> (Giddens, 2013, p. 37)

For Marx, the structural position of the worker was determined by their place within the value structure of capitalism. Marx (1986, p. 276) articulates this position with some clarity in Chapter 6 of Volume 1 of *Capital*, observing that what defines the wage worker under capitalism is that they have no other way to meet the costs of their subsistence except by selling their own labour power, which is the only thing of value that they can trade. The owner of capital, on the other hand, is only interested in the labour power of the worker because of

DOI: 10.4324/9781003257912-11

the difference between the wage that is paid for this labour and the value that can be extracted from the fruits of this expended labour power (Marx 1986, p. 279). Here, as Marx points out, there is a difference in use values: for the worker, the use value of selling labour power is that it is the only means of surviving; for the capitalist, labour power is only of interest because the surplus value can be attained through an exploitative relation. These two use values for labour power – between the worker and the owner of capital – are in conflict: capitalism rests upon this antagonistic 'dichotomous' relation.

The above summary of the theory of exploitation and value described by Marx, and its relation to the structure of classes, should be familiar enough and is one template for understanding the economic basis for class structure in contemporary human societies. However, what is less clear is how this value logic applies beyond the human. While there is scholarly contention on this issue (Foster and Clark, 2018), there is certainly a strong case that Marx reflected a hierarchical anthropocentrism in his understanding of capital. One area where we can perceive this anthropocentrism is in Marx's analysis of labour. It is relatively well known that Marx reserved the creative activity of productive labour for humans alone (Benton, 1993). We find stark evidence for this in the *1844 Economic and Philosophic Manuscripts* (Marx, 1978), where Marx differentiates the labour of animals from that of humans, arguing that while animals cannot distinguish themselves from their life activity, humans are purported to labour consciously and can produce regardless of necessity. This anthropocentrism lingers in late Marx too – for example, in Volume 2 of *Capital*, Marx levels an attack at Adam Smith, chastising him for comparing draught animal labour to human wage labour (Marx, 1992, p. 449 n 6; see also Luxemburg, 2003, p. 39).

To some extent, Marx's anthropocentrism prevented him from realizing the full implications of his analysis of capitalism as an economic and social force. In some respects, this represented a failure of Marx to apply his unique method of analysis with consistency. Luis Althusser (2015, pp. 268–70) argues that what made late Marx distinctive was his commitment to 'anti-humanism' – that is, his rejection of the humanistic values of nineteenth-century liberalism in favour of a frank analysis of the economic life of capital. While I would agree with this assessment by Althusser, we can certainly find examples where Marx's commitment to an anti-humanist approach falters – arguably his understanding of animals in production is one example. In these writings, Marx does not appear able to suspend his own anthropocentric viewpoint in order to effect a frank analysis of the relation of animals to production. While this is unfortunate, there is no reason why we cannot make use of Marx's approach to complete the analysis of capitalism. In this context, it is worth noting that recent social theory has offered correctives for how Marx's theory of value might move beyond the human. Notably, Jason W. Moore (2015) has developed an ecological reading of the history of capitalism and with it a revision of Marxist approaches to understanding capital and its relation to labour and nature. To some extent, this account resists the anthropocentrism inherent in Marx; echoing feminist and ecofeminist readings (Federici, 2014; Mies, 2014; Salleh, 2017), Moore seeks to highlight that the material base of capitalism does not solely come into being through human 'productive' labour, but that informal unpaid labour and non-human energies are also required to underpin value relations. Further, a range of other scholars have applied lessons from Marx's labour value theory to animals, including Donna Haraway's (2008, p. 46) articulations of 'lively capital' and 'encounter value'; Maan Barua's (2017) development of encounter value in his analysis of the politics of conservation; and Rosemary Collard and Jessica Dempsey's (2013) articulation of 'lively commodities' (see also Collard, 2020). All this work has provided glimpses of how Marx's value theory might have applied to animals if non-human productive activity were a focus of the volumes of *Capital*.

For sociology, one of the important outcomes of addressing the anthropocentrism within Marx's analysis of capital is to gain a stronger understanding of the place of animals within capitalism, and in particular, to improve our capacity to advance a structural analysis of animals and their relation to capital – that is, to understand the 'class' position of animals under capitalism. My aim in this chapter is to articulate the structural position of animals used for food under capitalism. As I will argue, following through Marx's value theory allows us to see that food animals exist as a unique hybrid of raw material, consumption commodity, and labour and that these beings comprise a large and distinctly important labour force of capitalism. Further, food animals as a social grouping are thrust into essentially structurally antagonistic relations, not only with humans but also – importantly – against fixed capital. The latter relation between food animals and fixed capital is exemplified in industrial animal agriculture, where animals confront enclosures, transport, and machines in a historically unprecedented encounter; for this reason, as the meeting point of extreme hierarchical anthropocentrism with the hyperbolic, rationalized overproduction of capitalism, the factory farm is central to producing food animals as a distinct economic class.

Raw material, consumption commodity, and labour

One of the central contradictions shaping human–animal relations is the use of animals by humans as raw materials in food supplies. This is a contradiction because prevailing knowledge systems continue to highlight that animals are sentient, have emotional lives, experience strong relational bonds and have an interest in their own existence; however, almost everywhere in the balance sheets of capitalist food systems, food animals are simply treated as inputs to production, which exist only to be transformed into commodities for a profit. In Marx's analysis of capital, a 'raw material' is essentially a resource that is an object of labour (Marx 1986, p. 285). As discussed above, Marx saw human labour alone as creative and transformative; in this view, labour works on a raw material, and through this work transforms it into a commodity with a new use value (Marx 1986, pp. 314–15). For Marx, it is this transformative capacity of labour that both allowed for the preservation of value in the object of labour and, through the expenditure of effort through work, allowed for the addition of new value, altering this object into something with a different use value. As I shall discuss below, there is no reason to imagine that only humans have this capacity; indeed, recognizing animal labour provides a glimpse into the quite distinct relation between animals and value production under capitalism. However, for the moment, it is enough to recognize that Marx's understanding of the 'raw material' describes at least one form that food animals take within animal agriculture: as a material that enters production as an input to be worked upon and transformed into a new use value (i.e. 'food').

Of course, animals involved in food production do not appear from nowhere; these raw materials themselves have to be 'produced'. Domestication implies control over reproduction; in the case of capitalist animal agriculture, this means forced reproduction on an industrial scale. For example, of the 75 million pigs in the United States, some 6.4 million sows are continuously 'farrowed' in order to produce litters of piglets, which will enable the creation of life to ensure the ongoing supply of animals who will become 'raw materials' for production (USDA, 2019). This control over reproduction is also central to dairy production, which relies on repeated forced insemination, pregnancy, and birth to enable continuous milk production (Gillespie, 2014, pp. 1326–27). The function of this mass-scale forced reproduction is to proliferate animals as raw materials for food production – that is, to generate living materials that will be used in production to become consumer commodities. This proliferation of animals as raw materials is significant. For example, in 2020 approximately 78 billion land animals were killed for human consumption, with chickens alone comprising 70 billion of these animals (UN FAOSTAT 2019); this means at any one moment globally there are approximately 40 billion living land animals within animal agriculture systems (UN FAOSTAT 2019). Consistent data is not available on fish within aquaculture, although one estimate suggests that up to 180 billion fish are living within these systems (Mood and Brooke, 2019).

The proliferation of animals as raw materials for food systems represents one aspect of the 'life' of animals under capitalism. The other prominent existence for animals is the transformation of their bodies, body parts, and bodily secretions into consumption commodities. Here, as described above, animals are worked on within production to enable a transformation into a product consumed by humans. For most of these animals, this production process necessitates the extinguishment of their life as part of this transformation; production represents the movement when they transition from living beings to dead meat. That this process is treated as 'valuable' highlights another contradiction and site of antagonism. Just as the human worker confronts a contradiction between the use value of their own labour to themselves (as a means of survival) and the use value placed on this labour by capitalism (as a means of profit), food animals also confront a contradiction, although it is of a different order: in order for animals as raw materials to be transformed into a consumption commodity, most animals must lose their lives – the existential source of all use values for them – in order to become a use value for capitalism as a consumption commodity.

The mass proliferation of animals as consumption commodities has had implications for animal life on a massive scale, but also marks a shift in the 'metabolic' relation between humans and animals under capitalism. The growth in animal-sourced foods has not shadowed human population growth in a parallel way; instead, the growth rate of animals as food has exceeded the human population growth rate. In 1961, global per capita meat consumption, excluding fish and seafood, was at 23 kilograms per person; by 2014, this had nearly doubled to 43 kilograms per person (Ritchie and Roser, 2017). World per capita fish consumption has more than doubled (UN FAO, 2018). For Raj Patel and Jason Moore, this expansion in the availability and consumption of animal-based foods is part of capitalism's strategy for creating 'cheap food' as a way to reduce the input costs of production: reducing the costs of animal-based food effectively reduced the wage cost of capitalism as a system, expanding the potential to extract surplus from labour (Patel and Moore, 2018). However, the mass availability of animal-based foods did not merely restructure the economic base of capitalism by altering the means of subsistence for human labour; it also realigned human and non-human populations in relation to each other, so that human survival has more and more overtly become premised upon the production and reproduction of animal lives through animal agriculture. In other words, on a structural level, capitalism fundamentally altered human–animal relations. Indeed, if we were to summarize this, we might note that under capitalism, animals have become more and more central to the reproduction of human life, and thus the 'social reproduction' of capitalism. Social reproduction is usually understood as comprising the forms of energy and labour required outside of formal production, which facilitate that production – for example, the forms of unpaid care labour that, while not formally 'priced', necessarily prop up the possibility of capitalism turning a profit. However, understanding the role of animals in value production highlights that food animals play a foundational role in human social reproduction. This means it is not just that animals participate in the 'social reproduction' of capital; rather, as animal-based food increasingly enables all forms of human production, including forms of social reproduction such as care work, we might observe that food animals under capitalism are increasingly a means for *the reproduction of reproduction itself*.

The existence of animals as both raw materials and consumption commodities within food systems is perhaps most visible to our contemporary knowledge systems: we most often 'see' animals in these commodity forms. However, it is the third aspect of the place of food animals in capitalist value chains that is perhaps the most important, even if it is less apparent: the existence of these animals as labourers. As we have discussed, Marx had an anthropocentric view of labour: the creative value-adding capacity of work was reserved for humans alone. However, recent work in animal studies and post-humanism has challenged the anthropocentrism of Marxist conceptualizations of labour by arguing that animals might well be considered labourers within production (e.g. Barua, 2017; Blattner, 2020; Cochrane, 2016; Collard and Dempsey, 2013; Coulter, 2016; Hribal, 2003; Noske, 1997; Painter, 2016; Perlo, 2002; Stuart, Schewe and Gunderson, 2013; Wadiwel, 2018, 2020).

We can push these analyses further by applying Marx's ideas of value to an analysis of food animal production. As discussed above, in the traditional Marxist view, labour exploitation occurs through capitalism seizing upon the difference between the value produced by labour power and the price of the wages paid to the worker (see Marx, 1986, pp. 270–80). In the archetypal labour relation described by Marx, the worker focuses their energies upon an object of labour, which is transformed by the work performed on it:

> While the productive labour is changing the means of production into constituent elements of a new product, their value undergo a 'transmigration' ['*Seelenwandrung*']. It deserts the consumed body to occupy the new created one. But this transmigration ['*Seelenwandrung*'] takes place, as it were, behind the back of the actual labour process. The worker is unable to add new labour, to create new value, without at the same time preserving old values, because the labour he [sic] adds must be of a specific useful kind, and he [sic] cannot do work of a useful kind without employing products as the means of production of a new product, and thereby transferring their value to the new product. The property therefore of which labour-power in action, living labour, possesses of preserving value, at the same time as it adds it, is a gift of nature which costs the worker nothing, but is very advantageous to the capitalist since it preserves the existing value of his [sic] capital.
>
> (Marx, 1986, pp. 314–15)

In this story, waged labour adds value by working on a raw material which is separated from their own bodies, sculpting and transforming this object of work. However, this narrative assumes that the object of labour is not coincident with the body of the labourer. Feminist approaches to labour have challenged this assumption, drawing attention to the way in which the metabolic processes of the workers themselves are drawn into production (Cooper, 2008; see also Beldo, 2017) – something we see in particular gendered labour practices, for example in commercial surrogacy work (Pande, 2014). This line of analysis proves highly useful for understanding food animal labour, which co-opts the biological processes of growth and reproduction in order to enable the transformation of a raw material into a finished product. These animals are required by production processes to work on themselves to produce new values. This labour, of course, is not waged; however, the same rationalities that drive capitalist economies, where the reproduction costs of labour are minimized in order to maximize output, prevail. Arguably, we can see these rationalities very clearly in play in attempts by animal agriculture to optimize 'feed conversion rates' (e.g. Pierozan et al., 2016), allowing industrial forms of animal agriculture to both expand 'yield' and potentially reduce the turnover time required before these animals are slaughtered to become a consumption commodity. Here, capitalism leaps upon the metabolic activity of animals as 'labour', extracting value by making use of the difference between the value produced by animals themselves in their own growth and 'feed conversion' activities, and the costs to the system associated with the reproduction of the lives of these animals.

We now arrive at a more complete picture of the structural location of food animals within the value logic of capitalism, one that points to the unique 'hybrid' position of these animals. Food animals represent a unique amalgam of raw materials, consumption commodities, and labour. In this sense, they are a hybrid combination of 'constant' and 'variable' capital, where constant capital represents an inert input to production and variable capital represents labour within the production system (Marx, 1986, p. 317). Food animals enter production as a raw material, with a use value as an input that will be 'value added' by the production process. These beings are compelled to perform metabolic labour upon themselves, 'trusted' by production to maintain the value that was invested in them as a raw material, and systematically add value to themselves as they transform into a new use value. Capitalist production has leapt on animals because of this capacity to produce value over and above the costs of their subsistence. Finally, these animals are transformed through a violent process of extraction, which in many cases requires the extinguishment of their lives, in order to effect a metamorphosis into a consumption commodity.

Over the last century, this commodity has served a globally important and accelerating function as a means of subsistence for human populations; this role means that this commodity has been increasingly positioned as underpinning, in existential terms, the reproduction of human life. While different human societies have historically made use of animals as a means of subsistence, arguably what marks the existence of food animals under capitalism as unique is their unprecedented function both as a source of surplus, driving the proliferation of animal-based foods and relatedly, their increasingly irreplaceable role as a means of subsistence, which underpins the productivity of the formal and informal human labour force; this means that, despite the growth in the consumption of plant-based foods globally, per capita animal-sourced food consumption continues to climb. It also means that as a social grouping, the animal labour force within food systems – particularly industrial land- and sea-based production – has grown substantially; as indicated above, it now perhaps totals more than 200 billion living beings. It is this unique constellation of factors that means, as I argue, that food animals emerge as a distinct economic class under capitalism.

Structural antagonism

As discussed above, for Marx class stratification was not merely defined by the standpoint of individuals with respect to the means of production, but also shaped by conflict and antagonism. In part, this antagonism is about differing use values for labour power: the human worker sells their labour to survive; capital, on the other hand, has no essential interest in the workers but is only interested in the value that can be extracted from the expenditure of this labour power. This, in turn, leads to another site of antagonism: labour time. As outlined in Chapter 10 of *Capital*, the working day represents a pitch battle between the interest of the worker to minimize their time labouring for capital, and the opposing drive of production to capture as much of the worker's time as possible and convert it into surplus (Marx, 1986, pp. 340–416). For some thinkers, such as Antonio Negri and Mario Tronti, this conflict is defining of the working class, who by default engage in an antagonistic resistance to the compulsion to labour (Negri, 1991; Tronti, 2019).

The antagonisms that shape food animal labour differ in character from those that shape the human wage worker. We can identify three different lines of this antagonism shaping the existence of food animals as an economic class. First, hierarchical anthropocentrism places animal interests at odds with those of humans. I have previously argued that the biopolitical violence that characterizes human–animal relations is essentially warlike in nature (Wadiwel, 2015). Most animals that are in direct relation with humans are subject to systematic control, violence, and domination directed at achieving satisfaction of human pleasures and interests. This relation is most apparent within intensive food systems, where the interests of humans in consuming animals as a means of subsistence and pleasure exist in direct confrontation with the interests and will of animals to be free from this violence. Industrial capitalism intensified this war, achieving an intoxicating and hyperbolic proliferation of violence against animals. Food animals are located at the heart of this antagonistic set of relations.

Second, the interaction between labour power and time is dramatically different for food animals compared with human workers; this in turn highlights heterogenous sites of antagonism. As discussed above, the human working class is caught in a tussle over labour time, with a continual drive by capital to transform all time into labour time. Marx understood this as the process of capital seeking to expand absolute and relative surplus value (Marx, 1986, p. 342). Expansions of absolute surplus value were possible by extending the working day – that is, compelling the human labour force to work increasingly longer hours. Meanwhile, alteration in relative surplus value was possible by intensifying the working day or compelling labour to produce more within the same amount of time. However, these dynamics differ completely when we examine food animals in production (Wadiwel, 2018, 2020). There is no working day for animals, since their whole day – indeed, their whole life – is completely subordinated to capital. This means that the working day cannot be expanded further to maximize absolute surplus value. However, other strategies are possible, and these have actively been utilized by capitalist animal agriculture to shape production. First, while the working day cannot be increased, the lives of animals can be shortened, reducing turnover times and thus increasing profitability. It is thus no accident that over the past 50 years broiler chickens have been genetically selected to effectively halve 'growing' time (Petracci et al., 2015, p. 364). Second, as discussed above, productivity can be enhanced by increasing 'yield' through improving 'feed conversion' rates – that is, making animals grow faster over shorter periods of time (Pierozan et al., 2016). The sum of these interactions is a different pattern of antagonism experienced by the human worker compared with the food animal. For the human worker, the site of contest is the working day: how much time they are compelled to work and the intensity of labour commanded during this period. For the food animal, this contest occurs on a different scale and is more overtly biopolitical and existential: surplus depends upon how long these animals will need to live in order to produce a 'finished product' and the intensity of processes involved, no matter how much suffering and discomfort they generate, to 'yield' a saleable profitable product. Animals, pushed to the barest possible lines of existence, want to avoid suffering and live; capitalist animal agriculture, stripped to its most basic rationalities, seeks the opposite goals in the name of enhancing surplus.

Finally, animals are placed into antagonistic relations, not merely against humans but also against forms of 'fixed capital': buildings, enclosures, tools, and machines. Inherent to capitalism is a continually altering 'technical composition', where the mass of fixed and circulating capital (raw materials, buildings, machines) increases over time as human labour is replaced and displaced from production (Marx, 1986, pp. 762–81). This has long been the relation between machines and human workers in production; machines progressively replace humans in order to enhance relative surplus value, leading in some cases to the full automation of production, with significant implications for human labour politics (Mason, 2016). However, the labour of food animals in global animal agriculture cannot be replaced by machines. This is because these animals, as discussed above, are raw material and labour, and *become the commodity*. This means the relation of animals to mechanization has differed from the relation between humans and machines under capitalism.

Within animal agriculture, humans are replaced with machines, with the goal of reducing human labour time. However, this evacuation of human labour is accompanied by an increase in the mass of animals in production, in order to increase the volume of commodities produced – that is, animal-based foods. The mass of animal labour time, therefore, increases as mechanization intensifies. This marks the difference between the structural relationship of technologies to humans and animals, respectively. For human labour, the arrival of machines signals a reduction in labour time and potentially the loss of work; for animals, the arrival of machines signals the disappearance of humans from production – sometimes, as in the case of robotic dairy farms, a complete evacuation – and the simultaneous massification of animals as raw materials, as labour, and as final commodities.

There is one more peculiarity that we should note. As animal agriculture intensifies – as machines, raw materials, buildings, enclosures, and so on progressively replace human labour – animals more overtly confront *fixed capital* rather than humans in a primary antagonistic relation. In summary, this is the broad outline of the factory farm, the perfect combination of hierarchical anthropocentrism and industrial capitalism: intensive animal agriculture represents the arrival of enclosures, buildings, and machines, the evacuation of human labour time, the massification of animals in production, and the replacement of the human–animal relation with a naked and brutal confrontation between animals and fixed capital. I stress this, noting that while the field of human–animal studies in recent times has primarily been interested in documenting relational encounters between humans and animals, the very real relation facing most animals in food systems is a hostile tussle with technologies, enclosures, and machines – that is, fixed capital. The response of billions of animals to this primary 'relation' is resistance. As I have discussed elsewhere, the technologies deployed in animal agriculture aim at enabling the subordination of animals to the imperatives of production; they work to counter and prevent resistance of animals and ensure compliance with the rhythms of production (Wadiwel, 2016, 2018; see also Hribal, 2003). This tussle between the resistance of animals and the continual attempts to use fixed capital to counter this insubordination marks the particular shape of the antagonism between food animals as a class and the forces of capital.

Conclusion

As discussed above, Giddens (2013, p. 37) suggests that Marx develops a theory of class that is 'basically dichotomous': that is there are two classes, human wage workers and the capital-owning bourgeoisie. In this chapter, I have argued that food animals constitute a distinct and differentiated economic class, and thus, against the tendency of Marx, I would suggest that it is not possible to have a dichotomous view of class relations. Food animals represent an enormous labour force, comprising more than 200 billion living beings. These animals have a distinct structural positioning as a hybrid of constant and variable capital, representing at different phases of production, raw material, consumption commodity, and labour. This economic class has its own antagonistic relations, which mark its distinct orientation *vis-à-vis* humans, labour time, and fixed capital. Humans have historically made use of animals as a means of subsistence in a variety of traditions and cultures; however, capitalism saw the emergence of food animals as a distinct social and economic group, marked by a unique relation to capital.

Describing food animals as an economic class is both enlightening and sobering. On one hand, this description helps to clarify the unique place of animals under capitalism and offers a corrective for Marxist theory, which has by and large neglected this analysis. However, the perspective offered in this chapter also starkly illustrates that we cannot assume any solidarity between human labour and animal labour, since the structural positions of these two labour forces are potentially antagonistic. For animal advocates, this picture is potentially challenging. Advocates must work to build solidarity, in particular, to counter the hierarchical anthropocentrism that is one of the trajectories that sustains animal agriculture. But there is also a challenge to work authentically to represent the interests of animals, which are articulated by their resistive relations. The insubordination of animals to capital is reflected in a continuing desire to escape and live a full life that is not dominated by capital in combination with hierarchical anthropocentrism. The challenge for animal advocates is to assist animals as a class to realize this goal.

References

Althusser, L. (2015). The object of capital. In L. Althusser, É. Balibar, R. Establet, P. Macherey and J. Rancière (eds), *Reading Capital: The Complete Edition*. London: Verso.

Barua, M. (2017). Non-human labour, encounter value, spectacular accumulation: The geographies of a lively commodity. *Transactions of the Institute of British Geographers*, 42(2), 274–288.

Benton, T. (1993). *Natural Relations: Ecology, Animal Rights, and Social Justice*. London: Verso.

Beldo, L. (2017). Metabolic labor: Broiler chickens and the exploitation of vitality. *Environmental Humanities*, 9(1), 108–128.

Blattner, C. (2020). Should animals have a right to work? Promises and pitfalls. *Animal Studies Journal*, 9(1), 32–92.

Cochrane, A. (2016). Labour rights for animals. In R. Garner and S. O'Sullivan (eds), *The Political Turn in Animal Ethics*. Lanham, MD: Rowman & Littlefield.

Cooper, M. (2008). *Life as Surplus: Biotechnology and Capitalism in the Neoliberal Era*. Seattle, WA: University of Washington Press.

Collard, R.-C. (2020). *Animal Traffic: Lively Capital in the Global Exotic Pet Trade*. Durham, NC: Duke University Press.

Collard, R.-C. and Dempsey, J. (2013). Life for sale? The politics of lively commodities. *Environment and Planning A*, 45(11), 2682–2699.

Coulter, K. (2016). *Animals, Work, and the Promise of Interspecies Solidarity*. New York: Palgrave Macmillan.

Federici, S. (2014). *Caliban and the Witch: Women, The Body and Primitive Accumulation*. New York: Autonomedia.

Foster, J.B., and Clark, B. (2018, 1 December). Marx and alienated speciesism. *Monthly Review: An Independent Socialist Magazine*. https://monthlyreview.org/2018/12/01/marx-and-alienated-speciesism

Giddens, A. (2013). *Capitalism and Modern Social Theory: An Analysis of the Writings of Marx, Durkheim and Max Weber*. Cambridge: Cambridge University Press.

Gillespie, K. (2014). Sexualized violence and the gendered commodification of the animal body in Pacific Northwest US dairy production. *Gender, Place & Culture*, 21(10), 1321–1337.

Haraway, D.J. (2008). *When Species Meet*. Minneapolis, MN: University of Minnesota Press.

Hribal, J. (2003). 'Animals are part of the working class': A challenge to labor history. *Labor History*, 44(4), 435–453.

Luxemburg, R. (2003). *The Accumulation of Capital*. London: Routledge.

Marx, K. (1978). 'Economic and Philosophic Manuscripts of 1844'. In R. Tucker (ed.), *The Marx-Engels Reader* (2nd ed.). New York: W.W. Norton.

Marx, K. (1986). *Capital, Vol. 1*. Harmondsworth: Penguin.

Marx, K. (1992). *Capital, Vol. 2.* Harmondsworth: Penguin.

Mason, P. (2016). *Post-capitalism: A Guide to our Future.* Harmondsworth: Penguin.

Mies, M. (2014). *Patriarchy and Accumulation on a World Scale: Women in the International Division of Labour.* London: NBN International.

Mood, A. and Brooke, P. (2019, June). Estimated numbers of farmed fishes living in global aquaculture. *Data Spreadsheet.* Fishcount.org.uk. http://fishcount.org.uk/fish-count-estimates-2/numbers-of-farmed-fish-slaughtered-each-year

Moore, J.W. (2015). *Capitalism in the Web of Life: Ecology and the Accumulation of Capital.* London: Verso.

Negri, A. (1991). *Marx Beyond Marx: Lessons on the Grundrisse.* New York: Autonomedia.

Noske, B. (1997). *Beyond Boundaries: Humans and Animals.* Montreal: Black Rose Books.

Painter, C. (2016). Non-human animals within contemporary capitalism: A Marxist account of non-human animal liberation. *Capital & Class,* 40(2), 1–19.

Pande, A. (2014). *Wombs in Labor: Transnational Commercial Surrogacy in India.* New York: Columbia University Press.

Patel, R. and Moore, J.W. (2018). *A History of the World in Seven Cheap Things: A Guide to Capitalism, Nature and the Future of the Planet.* Melbourne: Black Inc.

Perlo, K. (2002). Marxism and the underdog. *Society & Animals,* 10, 303–318.

Petracci, M. et al. (2015). Meat quality in fast-growing broiler chickens. *World's Poultry Science Journal,* 71(2), 363–374.

Pierozan, C.R. et al. (2016). Factors affecting the daily feed intake and feed conversion ratio of pigs in grow-finishing units: The case of a company. *Porcine Health Management,* 2(1). https://doi.org/10.1186/s40813-016-0023-4

Ritchie, H. and Roser, M. (2017, August). Meat and seafood production and consumption. *Our World in Data.* https://ourworldindata.org/meat-and-seafood-production-consumption#per-capita-milk-consumption

Salleh, A. (2017). *Ecofeminism as Politics: Nature, Marx and the Postmodern.* London: Zed Books.

Stuart, D., Schewe, R.L. and Gunderson, R. (2013). Extending social theory to farm animals: Addressing alienation in the dairy sector. *Sociologia Ruralis,* 53(2), 201–222.

Tronti, M. (2019). *Workers and Capital.* London: Verso.

UN FAO (2018). Is the planet approaching 'peak fish'? Not so fast, study says. www.fao.org/news/story/en/item/1144274/icode

UN FAOSTAT (2019). www.fao.org/faostat/en

USDA (2019, 27 June). Quarterly hogs and pigs. https://downloads.usda.library.cornell.edu/usda-esmis/files/rj430453j/3b591k937/5m60r2937/hgpg0619.pdf

Wadiwel, D. (2015). *The War against Animals.* Leiden: Brill.

Wadiwel, D. (2016). Do fish resist? *Cultural Studies Review,* 22(1), 196–242.

Wadiwel, D. (2018). Chicken harvesting machine: Animal labor, resistance, and the time of production. *South Atlantic Quarterly,* 117(3), 527–549.

Wadiwel, D. (2020). The working day: Animals, capitalism and surplus time. In C.A. Blattner, K. Coulter and W. Kymlicka (eds), *Animal Labour: A New Frontier of Interspecies Justice?* Oxford: Oxford University Press.

8 Resisting *Zoopolis*

Bordering species relations as a response to COVID-19

Erika Cudworth

Introduction

Zoonotic pandemics originate in non-human animals, with microbial agents leaping from animal to human bodies (Quammen, 2012). They shine an uncomfortable light on the ways in which how we live, and in particular how we eat, have facilitated the sharing of pathogens across species. This chapter argues that the causes of zoonotic disease, which emerge from human engagement and misuse of wildlife and habitats, and from the use of animals as food, have not been addressed in policy responses which have largely focused on public health and the medical management of infection. The chapter understands the predominant policy responses to the zoonotic pandemic in terms of bordering and rebordering practices – restricting travel and mobility for some humans, attempting to limit trades in live animals and curtailment and surveillance of everyday practices. What COVID-19 illustrates, however, is that bordering – whether constituted by state security of territory or cultural, economic and social activities – is a process that reveals the extent of human bioinsecurity. In times of zoonotic pandemic, Foucault's (1991) notion of biopolitics needs to extend to the idea of zoopolitics to include the non-human lifeworld as an object of power (Shukin, 2009); and in pandemic times, the idea that politics is post-human rather than a discrete sphere of human activity gains traction. However, rather than opening up to the critical voices demanding a different kind of relationship with non-human animals, the awareness of rapidly developing new strains of pandemic disease has given salience to zoonotic discourses of bordering.

This chapter begins by considering the ways the current pandemic reflects, and exacerbates, intersected forms of inequality, including those of species. It then examines the relationship between zoonotic pandemics and the human treatment of other animals and the living world more broadly. A brief consideration of the current policy vacillation around the new/old normal precedes an exploration of bordering theory as a way of understanding the dominant responses to COVID-19. The final section considers the inherent problems of such responses in the context of the material entanglements of species and argues for post-humanist politics of species reordering.

DOI: 10.4324/9781003257912-12

COVID-19 and intersected social inequalities

Infectious diseases have been understood to be embedded in complex forms of inequality (Farmer, 2001), and COVID-19 has been no different in exacerbating existing forms of inequality (BBC, 2020b). Writing for the International Monetary Fund (IMF), Joseph Stiglitz (2020) contended that:

> COVID-19 has not been an equal opportunity virus: it goes after people in poor health and those whose daily lives expose them to greater contact with others. And this means it goes disproportionately after the poor, especially in poor countries and in advanced economies like the United States where access to health care is not guaranteed.

Data from the United States shows significantly higher rates of infection and death for Black, Hispanic/Latinx and Indigenous communities (Romano, 2020, p. 44). In the United Kingdom, the mortality rate among those hospitalized from ethnic minority groups is twice that of White British patients, and structural racism has been held partly responsible, with ethnic disparities in COVID-19 understood as part of a historical trend of poorer health outcomes for marginalized groups (Razai et al., 2021, p. 4). COVID-19 has also had important implications for gender terms. There is evidence, for example, that national and local lockdowns, lack of work and homeworking have led to a significant rise in gender-based violence in the home. This is combined with a withdrawal of government support for services and has led to the notion of 'disaster patriarchy' as male domination reasserts itself (V, 2021). Naomi Klein's (2007, 2017) notion of 'disaster capitalism' similarly explains how public disorientation in response to a collective shock enabled governments to push through radical pro-corporate measures. Klein has described COVID-19 as a 'perfect example' of a disaster that facilitates 'shock politics' (Solis, 2020).

Disasters do not just reflect and exacerbate intra-human forms of hierarchy; they powerfully inscribe humancentrism. The lives of animal companions are subject to extreme precarity in situations of emergency – for example, in the aftermath of Hurricane Katrina in the United States, 'rescuers forced people to leave their companion animals. Residents faced the choice between leaving animals they considered family members and risking their own lives' (Irvine, 2006, p. 4). In such circumstances, the lives of millions of farmed animals are seen as expendable. Confinement feeding operations (intensive industrial farming units) offer no chance for escape from flood, fire or structural damage. In addition, as Irvine (2006, p. 13) points out, 'the farmers who feed the animals do so by contract with large corporations who manage dozens of production facilities. Because the farmers do not own the animals, they cannot legally authorize or conduct rescue operations'. In any case, the 'rescue' of such creatures would cost more than their short lives are deemed to be worth.

The COVID-19 disaster is also an example of 'disaster anthroparchy', both reflecting relations of human domination and offering new avenues for the intensification of violence against non-human animals and confirming their expendability. As might be expected, the treatment of farmed animals in some countries was even more grim during the pandemic than the everyday routinized mass violence that characterizes animal agriculture (Cudworth, 2015). For example, in the United States between the end of April and mid-September 2020, pigs and chickens were subject to 'depopulation' by alternative methods deemed acceptable while slaughterhouses were closed, but which have been identified as highly unethical in causing prolonged suffering. Two million 'meat chickens' and 61,000 'laying hens' were killed by methods including smothering with foam (such as is used in fire-fighting). Up to 10,069,000 pigs were likely to have been killed over the summer of 2020, by various methods including ingesting poisoned food, being suffocated by the closing of ventilators and being subject to 'blunt force trauma' – meaning, for example, piglets being thrown to the ground until they are dead (*The Guardian*, 2020). Other creatures have been victims of the economic disruption caused by the current crisis – 'racing' animals such as horses and greyhounds, animals confined in laboratories, zoos or 'wildlife parks' – and subjected to a culling spree. As Paula Arcari (2020) remarks:

> When things go to shit, animals are on their own, which is what makes their entrapment in capitalist political economies so doubly heartless. That this animal-industrial complex is so directly implicated in the COVID-19 pandemic and the climate crisis, with myriad animals being substantial victims of both, only emphasises the cycles of violence that result from capitalist commodification.

The COVID-19 pandemic exposed the fragility of current systems of social organization, which exclude, consume and oppress, but it also provided avenues for the intensification of the way in which those relational systems of oppression routinely operate. Francis Nyamnjoh (2020, p. 8) suggests that COVID-19 operates as a prism through which various intersected inequalities might be examined. These inequalities, however, need to include those of species hierarchy.

Zoonotic pandemics and species relations

Animal abuse and environmental issues are linked and give rise to major public health issues. The World Health Organization (2010) has for some time suggested that the live animal trade, eating animals and industrialized animal agriculture have combined to generate zoonoses. More recently, for the United Nations Environment Programme, the COVID-19 pandemic was described as a 'warning that nature can take no more' and that 'humanity's destruction of nature' must stop (Andersen, 2020). Conservationist and primatologist Jane Goodall has claimed that the pandemic is directly related to human treatment of animals: 'Our disrespect for wild animals and our disrespect for farmed animals has created this situation' (Goodall, cited in Harvey, 2020).

Zoonoses are linked both to eating animals and encroaching on their spaces. Zoonotic diseases 'jump' from animals to humans, with the likelihood of this happening increasing when habitats of wild animals are disrupted and animals are farmed intensively. Human settlement, work, transport and a range of social practices make the lives of vulnerable creatures more so, encroaching on and eliminating habitats, and driving wild animals into closer proximity with humans (Quammen, 2012). Pathogenic jumping can involve moving through a number of species, such as the case of the 'Spanish influenza of 1918–1919, which had its ultimate source in a wild aquatic bird and, after passing through some combination of domesticated animals [...] managed to kill as many as 50 million people' (Quammen, 2012, p. 20). Ebola originates in bats and has decimated chimpanzee and gorilla populations as well as killing humans (Aguirre, 2017), while HIV-1 emerged from chimpanzees and has killed around 30 million people worldwide (Quammen, 2012, p. 41). Avian flu emerged in Hong Kong and originated in poultry, while SARS was traced to horseshoe bats (Daszak, 2020). MERS is associated with dromedary camels (Ji, 2020) and is far deadlier to humans than COVID-19. MERS has been flagged as the possible 'next' pandemic, particularly as desertification associated with climate change has meant a shift from cattle to camels in parts of Africa, China and Mongolia in addition to high camel populations in the Middle East (Kushner, 2021).

COVID-19 highlights the way our dysfunctional and exploitative 'relationships with animals are [not only] driving the emergence of zoonotic infectious diseases' (van Dooren, 2020; see also Jordan and Howard, 2020), but also destroying our life support systems. The scale of the loss of biodiversity is alarming. While human and domestic animal populations increase, wildlife populations plummet to such an extent that by 2018, 60 per cent of all mammals on Earth were farmed animals, 36 per cent were humans and just 4 per cent were wild animals; 70 per cent of the global bird population was farmed (Bar-On, Phillips and Milo, 2018). We are now experiencing the sixth mass extinction event (Kolbert, 2014), with forecasts that a million non-domesticated animal species will become extinct in the next few decades (IPBES, 2020).

It seems that the emergence of COVID-19 is likely to have been related to wildlife crime, which enabled the selling of wild animals which are protected species under international agreements, for food at 'wet' or fresh food markets in China (Xiao, Newman and Buesching, 2021). As a result, the 'wet markets' of Asian countries were misrepresented (Fine Maron, 2020; Westcott and Wang, 2020) in an attempt to assert that this zoonotic epidemic was an isolated incident. However, the conditions enabling zoonotic disease, as previous outbreaks of bird and swine flu have shown, are an endemic condition of the global networks of commoditization in the industrial systems, which turn non-human creatures into food (Rohr et al., 2019). The speed with which zoonoses are emerging challenges us to rethink those practices that encourage their emergence and how animals – particularly wild and farmed animals – are incorporated into social relations.

The new/old normal

As was demonstrated during the COVID-19 pandemic, the short-term policy frames of capitalist governments are undone by unexpected events in an increasingly unpredictable world. These policy frames are inadequate when we need to think not only about responsibilities to the next generation of humans in a particular place but also about the complex vulnerabilities we may cause for generations yet to come, and in different parts of the globe, alongside our situation in webs of relations with multifarious non-human species. The 'old normal' is an era in which mass human poverty abounds, where catastrophic climatic change threatens life on the planet and in which some species are already experiencing an extinction crisis. Yet, this is the normal to which neo-liberal (and other) governments quickly encouraged businesses, workers and public sector services to return.

The vast majority of global businesses (with the exception of industries and services deemed essential) were ordered to physically close in the vast majority of countries once a global pandemic was confirmed by the World Health Organization, with rare exceptions such as Sweden and Hong Kong. From mid-March 2020, in the first national lockdown in the United Kingdom, schools and universities were shut, offices deserted and most shops closed. Staff in some organizations were furloughed and the United Kingdom witnessed a mass shift to working from home (Parry et al., 2021). The 'environment' received a boost in the form of a minor and short-lived decline in carbon emissions – the news media reported an explosion of wildlife activity as flights were grounded and roads were quiet. Fewer hedgehogs, for example, were killed by traffic on the roads of Poland (Łopucki et al., 2021), while reduced shipping and ocean travel reduced the risk to marine mammals and the disruption to marine life from ocean noise pollution such as fishing sonar (Livermore, 2020). Yet, many of the immediate positive effects of the pandemic on wildlife – such as reduced road, air and ship deaths or disruption – are likely to have been lost with the return of business as usual. The journal *Nature* reports that while CO_2 emissions dropped during the early phases of the pandemic, these were less than expected and are now being reversed (Tollefson, 2021).

When it comes to domesticated animals, there was an overwhelmingly negative impact, as we have seen above with respect to the worsening treatment of farmed animals. In Wuhan, where the virus was seen to emerge, the first months of the outbreak saw mass abandonment of animal companions as people were evacuated and were unable to return, leaving animals to starve, while in other Chinese cities, animal companions seen on the street were culled (Kim, 2020). Domestic animal rescue centres in the United Kingdom were inundated with animals abandoned due to job loss and financial strain, but particularly in the early phases of lockdown – reflecting the imagined 'threat' of contamination, particularly from cats, whose movements are less controlled than dogs (Battersea, 2020). Evidence suggests, however, that it is animal companions and not their 'owners' who are more vulnerable, with dogs and cats being shown to have caught COVID-19 from their human housemates (DEFRA, 2020; Sit et al., 2020).

By the first week of August 2020, the then UK Prime Minister, Boris Johnson, was encouraging people to return to work and by the end of the month there was a concerted government drive to get people back to work to promote productivity and shore up the economies of city centres (BBC, 2020a). The UK Chancellor of the Exchequer, Rishi Sunak, encouraged people to go out rather than stay at home, sponsoring cheap drinks and meals to support the hospitality industry (Hutton, 2020). The subsequent and predicted spikes in cases prompted the government to impose restrictions on socializing to no more than six people, and autumn 2020 saw increasing restrictions followed by a second national lockdown (BBC, 2020b). The vagaries of lockdown also saw COVID-19 outbreaks at meat processing factories, which prompted localized actions in European countries to increase worker safety and hygiene (Middleton, Reinjes and Lopez, 2020). In the summers of 2020 and 2021, international tensions attended the adding and removing of various countries from lists restricting travel to and return from, while the travel and aviation industries pushed for relaxing restrictions.

These fast-paced changes in policy direction reflect the tensions of bordering within capitalist economies. In many countries across the globe, people were divided from each other – through social distancing, being apart yet together in divided workspaces, through restrictions on travel and socializing, by avoiding public space and 'staying in and staying home'. We became subject to border cultures through these attempts to retain business as usual while shoring up our vulnerabilities to the pathogens of other animals – human and not. The tensions between capitalism and public health and environmental and planetary health were thrown into sharp relief. We also attempted to shore up borders during times of pandemic in particularly violent ways. The two sections that follow consider bordering as a response to COVID-19, particularly in relation to the separation of species (human and other), and the difficulties bordering politics encounter in ignoring or resisting zoopolis. Both are illustrations of the ways in which the so-called 'new normal' looks very familiar indeed.

Bordering as a response to COVID-19

COVID-19 has overwhelmingly been understood as a medical issue, the containment of which relies on advances in biomedical research and clinical treatment and public health measures (Oberlander, 2020). Many of those health measures constitute forms of erecting borders and policing these (from restrictions on national and international travel to restrictions on bodies in homes and public spaces) and on the maintenance of barriers (through social distancing and wearing face-coverings, for example). The international and intranational competition for the creation, patenting, testing and rolling out of vaccines can also be understood in terms of the maintenance and policing of borders.

Bordering is a relatively new and expanding field of scholarship influenced by decolonial theorists, international relations scholarship and migration and gender studies. Bordering theory is constituted by key concepts that include power, inferiority, epistemic difference, hierarchization and differential inclusion, which makes it fertile ground for scholarship in sociological animal studies. The early work of Gloria Anzaldúa (1999) examines the United States–Mexico physio-political border, where she roots her conceptualization of 'border culture'. Anzaldúa (1999, 2015) sees borderlands as most likely to be characterized by *la mezcla*, mixing or hybridity. This is in tension with the key function of the border, which operates to separate 'safe' and 'unsafe' and distinguish between 'us' and 'them' through a dividing line, while powerful discourses operate to demarcate legitimate inhabitants versus the illegitimate inhabitants along racialized lines. Mignolo and Tlostanova (2006) develop such ideas by situating 'border thinking' within both the geographic and epistemic discourses. They draw on colonial and earlier histories, where frontiers or borders were deployed as means of distinguishing civilization (and the civilized) from barbarism (and the barbaric people, fauna and flora of 'wilderness').

Discourses regarding the control of national borders have been central to political projects, but for bordering theory the process of bordering is not only political. State borders are only one kind of border among many (Nail, 2016). Rather, processes of bordering weave together social, political and economic configurations in complex ways (Cassidy, Yuval-Davis and Wemyss, 2018, p. 3). Nash and Reid (2010) assert that state bordering processes are double edged – related on the one hand to state territorial boundaries and on the other to symbolic social and cultural lines of inclusion and difference. Border thinking goes beyond physical-bounded spaces into the dynamics of exclusion and inclusion. Mezzadra and Neilson speak of these shifts as 'differential inclusion', which entails varying levels of acceptance, segmentation, discrimination and subordination within the same space (Mezzadra and Neilson, 2012). Narratives of the border(land) remain powerful, in spite of a complex array of de- and rebordering processes. In *Bordering*, Cassidy, Yuval-Davis and Wemyss (2018) examine the ways in which we can see borderwork taking place in everyday life and practices, and how this is contested, particularly by the human inhabitants of the borderland. They convincingly evidence that the deterritorialization of borders has seen their relocation in a multiplicity of spaces spread throughout civil society. Thus, borders are mobile (Sassen, 2013), paradoxical and invisible, yet impactful (Kolossov and Scott, 2013).

During COVID-19, we saw various attempts to border humans: restrictions on travel and movement, exclusions due to closures of schools, shops and so on and practices of biosecurity – from handwashing and mask-wearing to testing and tracing schemes. In homes of wealthy countries, during periods of national lockdown, the home space was subject to bordering and rebordering as the home became a different kind of territory for homeworking and homeschooling (Iossifova, 2020). The bodily bordering of social distancing and facial covering are preventative measures shaped by public health discourse so that rather than seeking to prevent something from happening, measures are designed to control the spread of the disease and targeted at humans only (Romano, 2020, pp. 35–37). Yet, the borders and frontiers of bordering theory are highly apposite when it comes to considering the relations implicating other species in the context of COVID. While zoonoses demonstrate our kinship with other animals (Shukin, 2009), the predominant policy response is not to rethink the problems attending our treatment of other creatures. Rather, separation – from both humans and other animals – is proposed in the interests of protecting humans from pathogens originating in animals and gaining 'mastery' over zoonotic diseases (Hinchliffe, 2015, p. 29). A human-centred managerialism has characterized responses. While a food market in China was identified as the apparent origin of the current pandemic, interrogation of the specific problematic issues of species encounters has so far been avoided.

I have already suggested that practices of inappropriate mixing – such as wild and domestic and live and dead animals in the wet markets of China – have been demonized. In the Western imaginary, such places and spaces of species mixing are associated with the racialized Other (Shukin, 2009). Thus, wet markets in China and bushmeat hunting in Central Africa (Wolfe et al., 2005) have been singled out as sources of zoonotic disease, along with the small-scale, mixed farming practised in much of the Global South. The practices that constitute them are seen as 'backward' and belonging to traditional societies that have no place in a modern, globalized world (Charles, 2020). Yet, the bordering practices that set these practices as exclusive to othered peoples and places are false, ignoring the wild animal trade, and the ways in which 'bushmeat' and food originating in wet markets are implicated in global food supply chains (Friant, Paige and Goldberg, 2015).

Wild animals are increasingly subject to bordering by being unable to move due to the destruction of their habitat through deforestation by logging, mining and other extractive practices, the spread of urbanization and the creation of new transport links for humans (Jones et al., 2008). Land-use changes affect wildlife species themselves and impact the health of both human and non-human animals (Ellis, McWhorter and Maron, 2012). The relentless expansion of human populations and dwelling space has had a catastrophic impact on 'wild' animals, with their populations dropping by an alarming one-third between 1970 and 2010; terrestrial and marine species declined by 39 per cent and freshwater species declined by an average of 76 per cent (World Wildlife Fund, 2021). This extermination by settlement has parallels with the impact of colonizing settler populations on Indigenous peoples. The bordering of habitat destruction contributes significantly to the emergence of zoonotic diseases because it fragments habitats, thereby increasing human contact with wild animals, as people or animals move in on each other's spaces (Jones et al., 2013). This ensures that wild animals who are under stress come into greater contact with humans and domestic animals, creating ideal conditions for the emergence of zoonoses (Seltmann et al., 2017). As David Quammen (2020) describes:

> We invade tropical forests and other wild landscapes, which harbor so many species of animals and plants – and within those creatures, so many unknown viruses. We cut the trees; we kill the animals or cage them and send them to markets. We disrupt ecosystems, and we shake viruses loose from their natural hosts. When that happens, they need a new host. Often, we are it.

Rather than rethink our attitude to other species and spaces, when it comes to the politics of zoonotic disease, separation and strict biosecurity are seen by health-related international organizations, such as the World Health Organization, as a means of preventing or stopping outbreaks at source (Hinchliffe, 2015, p. 29). Such practices of biosecurity operate in the intensive farming systems characterizing the Global North. Steve Hinchliffe argues that this sets the 'virtuous' bio-secure Global North against the 'interspecies intimacy' and 'contingency' characterizing the Global South (Hinchliffe, 2015, p. 30); thus, we have a bordering between the civilized practices of the West and the more intimate and leaky species boundaries in the spaces of Africa and Asia.

Bordering practices are also seen when we consider the search for animal hosts and the idea that 'singular animal species' bear 'the burden of epidemic blame' (Keck and Lynteris, 2018, p. 7). An ultimate form of bordering is killing, and one response to zoonotic disease is to extinguish members of the host species. Millions of farmed animals are killed, supposedly in the interest of protecting human health, in a zoonotic outbreak. For example, all the chickens in Hong Kong – over 1.2 million animals – were killed in response to the 1997 avian flu outbreak (Quammen, 2012). In November 2020, the Danish government rushed through a specious plan to cull up to 17 million farmed minks due to concerns that a mutated form of COVID-19 found on mink farms could hamper the impact of a vaccine for humans (Murray, 2020). This is an imperial human response that suggests the lives of animals are expendable and 'portraying animals as incubators, carriers, reservoirs, or spreaders of human infection […] grounds the scientific study of zoonosis on hard anthropocentric ground' (Keck and Lynteris, 2018, p. 10). Certainly, the science suggests that such culling is 'unwarranted vilification' of a host species (Xiao, Newman and Buesching, 2021). In our treatment of other animals during COVID-19, predominant policy reactions have been to attempt to shore up various kinds of borders: of individual humans and their kin; between species and between objects, spaces and territories. When it comes to the treatment of non-human animals, those bordering processes are particularly violent.

Zoopolis and the leaky boundaries of species

The politics of biopower as a technology centred on life can clearly be seen in the COVID-19 crisis (Foucault, 1991) – human population patterns, characteristics and behaviour are embedded in decision-making and the practice of daily life is governed by new regimes of bodily discipline such as handwashing and sanitizing (see Foucault, 1991). Yet, COVID-19 is also a powerful illustration of the problem of human-centric bias in Foucault's original formulation. There is a need for an extension of bio to zoo – as Nicole Shukin (2009) suggests. For Shukin (2009, p. 28), there is an 'inescapable bleed' between human social life and the lives of other species, but this is denied by a politics of pandemic biosecurity.

Yuval-Davis and Stoetzler (2002) argue that borders are fixed only in our imaginations; in fact, they are characterized by plasticity and permeability. This is why much of the bordering literature suggests that attempts – often violent and highly exclusionary – to shore up borders cannot secure populations on the 'inside'. Rather, borders are porous and constantly subject to change. In relations with other species, bordering practices are constantly breached and shown to be shaped by the tension between exclusions and expulsions on the one hand, and leakage and entanglement on the other. In their book *Zoopolis*, Donaldson and Kymlicka (2011) consider that the historical development of our relationships with other creatures has been inherently exploitative, but argue that huge varieties of species exhibit complex relations of interactivity with humans. Even the most 'wild' of wild animals do not live in pristine wilderness but are entangled with non-wild species and with humans at least to some degree. As a result, a 'dizzying array of relationships' exist with diverse origins, different spatial dimensions and levels of dependency, interaction and vulnerability (Donaldson and Kymlicka, 2011, p. 68). This makes our attempts to border species highly problematic.

In thinking about transforming our currently violent relationship with animals, Donaldson and Kymlicka (2011) extend established political notions of sovereignty, justice, rights and citizenship in order to add animals to the political community or zoopolis. Wild creatures may share human territory but not our society, are not dependent on humans and might best be understood as 'sovereign communities'. Such populations are vulnerable to the impacts of human activity, however, through human encroachment on animal territory and associated habitat loss; through direct violence (hunting, trapping, culling, fishing) and due to 'spill over harms', which include pollution and climate change (Donaldson and Kymlicka, 2011 pp. 156–57). Donaldson and Kymlicka (2011, p.168) argue that we have positive duties to promote the flourishing of these populations in ways that account for ecological viability, multidimensionality and sharing of territory, human and animal mobility and the possibilities for sustainable cohabitation (Donaldson and Kymlicka, 2011, p. 191). A second category of 'liminal animals' comprises creatures not directly dependent on humans but on 'human settlement', in that they have made our habitats their own yet are often invisible to humans, or subject to culling as 'pests' (Donaldson and Kymlicka, 2011, p. 212). Some of these animals may be urbanized 'wild' species, opportunist species, escaped exotics or feral domesticates. Donaldson and Kymlicka (2011, p. 75) suggest that such creatures have a legitimate presence and require greater protection as 'denizen' communities. Domesticated animals are a very different case. While the original purpose of domestication was the use of animals to serve human ends (food, clothing, transportation and security), not all domesticated relationships necessarily mean the instrumentalization of animals, and dependency does not necessarily involve abuse and domination (Donaldson and Kymlicka, 2011, p. 75). Donaldson and Kymlicka (2011, p. 101) argue that domesticated animals should be understood as citizens on the grounds that they have 'belonging' (they live with us and share our lives); that their interests count in determining the 'public good' and that they have agency and should be able to shape the rules under which they live.

Ultimately, if the recommendations of *Zoopolis* were adopted, it is unlikely that a pandemic such as COVID-19 would be possible. Animals would not be farmed or eaten, wild animals and their habitats would be given a range of protections, domesticated animal populations would be hugely reduced and live as citizens and liminal populations living alongside humans (such as bats) would be respected rather than culled or displaced. The context of COVID-19 illustrates our bodied, vulnerable condition and our precarity as human animals particularly vividly. If our task is to try to learn to live and die well together, in our troubled present on a damaged Earth, then we need to consider what relations we might choose to make or retain and what we should also choose to sever in trying to promote multispecies flourishing. Rejecting a culture that equates the eating of animals with food would seem a relatively painless way for humans, particularly in the wealthy world, to live less wrongly. Concerted action to reverse the tide of industrial animal farming is crucial, as the avoidance of meat and dairy products is being hailed as the 'single biggest way' to reduce human impact on the environment (Poore and Nemek, 2018). The necessary challenge of COVID-19 is to force us to rethink the exploitation of other animals that underpins human societies in different ways, and to different extents, across the globe.

Yet, there are two key difficulties with the recommendations of *Zoopolis*. First, it is very unclear how any of this might be realized in societies that are systematically organized in terms of interlocking oppressive hierarchies of difference (of species, gender, ethnicity and more) and the resource-extractive exploitations of the Capitalocene. Second, the human remains the architect of the possible lives of other creatures and this is not subjected to scrutiny in *Zoopolis*. This is insufficiently radical – in reimaging species and reordering species relations, reimagining what it means to be human is crucial.

In moving away from the liberal humanism that undergirds Donaldson and Kymlicka's (2011) work, post-humanism can help us to think about the qualitative and quantitative shifts needed 'in our thinking about what exactly is the basic unit of common reference for our species, our polity and our relationship to the other inhabitants of this planet' (Braidotti, 2013, p. 2). There are many post-humanisms, however. My own use of the term allies with critical post-humanism which certainly should not be confused with the idea of a world *after* humans. In other words, discussions about the extinction of the species that might follow perhaps from a major climatic breakdown and/or system-wide conflict. It should also not be confused with transhumanism which, for example, focuses on body modification and/or uploading consciousnesses to computers – this is a hyper-humanism. Rather, a key feature of post-humanism is the rejection of the *humanist idea of the human* as the measure of everything else. Humanism, as a mode of thinking, replaced religious authority as the ultimate source of knowledge with human science as the final arbiter. In doing so, humanism put the human at the centre as a species that could be considered as 'essential' and 'unique' – as separate from the rest of nature. This led to two highly problematic outcomes which are of utmost importance in thinking about zoonotic pandemics. First, it led to the view that the rest of nature could be known and as a result controlled. Second, it framed non-human nature as resources that humans could draw upon for their own ends. Post-humanists reject this central separation of the human species as separate from other species and the rest of nature. Hence, a prime feature of post-humanist work has been to de-centre the human. A critical post-humanist position represents a reaction against the view of human exceptionalism which understands humanity to be marked off from the huge diversity of non-human animal life due to apparently exceptional characteristics. One of the key aims of post-humanist thinking has been to call for a radical re-thinking of the basis of the human relationship with the rest of nature. This is in part to reflect the character of 'inter species dependencies' and in part based on the self-interested need to reorient our relationship with the rest of nature in order for our species to survive. This means a radical rethinking also of our relationships with humans and of what it means to *be* human.

As post-humanism understands relationships between different forms of life as co-constituted and interdependent, the politics of bordering can be understood to represent a denial of the co-constitution of the social and the natural. Such denial, for example, has been a key element of particular kinds of approaches to rewilding. Jørgensen (2015) argues that a notion of rewilding premised on the dissociation of humans from nature dangerously reproduces nature–culture binaries. This is not to suggest rewilding projects are necessarily problematic – certainly enabling increased autonomy of non-human nature can be positive (Woods, 2005, p.177). However, rewilding projects that disavow human intrusion are ultimately experiments with an ideal of 'wilderness' that may place various animals, (including) humans and flora at risk. They are also framed by a deeply colonial approach to the 'management' of 'wilderness' (Ward, 2019). Most European projects therefore do not exclude humans, and understand rewilding as necessarily focused on coexistence and cohabitation (Prior & Ward, 2016). Such a move away from a

policy based on an understanding of separation, rather than change necessi-
tated by co-constitution, has much to inform responses to pandemic risk.

To understand relationships through a post-humanist lens which foregrounds
co-constitution and eschews separation does not undermine the analysis of the
operation of domination and exploitative power. In my work independently
and with Steve Hobden, I have sought to demonstrate not only the interdepen-
dence and co-constitution of human social life with other creatures and non-
human nature but also the ways in which human relations with the former are
bound up with human social systems of exclusion and oppression – of relations
of capital, coloniality and patriarchy (see Cudworth, 2011; Cudworth &
Hobden, 2011, 2018). From a post-humanist perspective, boundaries are inevitably leaky. Post-humanist critique raises vital questions for human beings in
the world and demands a more profound shift. While *Zoopolis* envisages radical
changes in relations with various animal species, it does not account for the co-
constitution of the world or of the ways exploitative practices are never exclusively human. Reducing the risk to humans of zoonotic disease requires an
understanding of neoliberal capitalist and imperialist imperatives that fuel species extinction, habitat destruction and climate forcing (see Malm, 2020). We
live in no separate history or present, rather, it is the assumption of borders that
drives a misunderstanding of the causes of a zoonotic pandemic. The idea that
the 'wild' can be secured by separation from 'humans' is a false premise, framed
by Western colonialist assumptions. The notion that zoonotic pandemics are
best managed by a public health response and a biomedical model of management alone, rather than understanding that we are interdependent, attached
and implicated in/as nature is a strategy that refuses to understand the fragility
of the world and the place of humans within it.

Conclusion

Our lack of acknowledgement of the value of the lives of other animals that
we raise or hunt to kill and eat, whose habitats we destroy and encroach upon,
whose populations we squeeze to the point of breaking, has led us to the current situation of crisis in human health and wellbeing. The worlds of high politics are so resistant to recognizing this that it is no wonder that engaging with
the enviro-socio-politico-economic causes of COVID-19 has been avoided
thus far. The pandemic was tackled by a humanocentric sticking plaster
approach to public health based on bordering, which failed to appreciate and
tackle the causes of zoonotic spillover and link this to extractive and exploitative systems of social practices. The order of species needs to be challenged.
Our relations with other species are co-constituted with intra-human exploitations, exclusions and vulnerabilities that are tied in with ways of living involving animal killing, displacement and exploitation. Challenging this would be a
radical change indeed.

Yet, a further move is required. The very notion of species is bound up
with the Western idea of what it means to be human and is gendered,

racialized and colonial; such categories are leaky too. In making the case for respect and response-ability (to follow Haraway, 2016), we need a situated perspective that takes account of different kinds of relations and possibilities, rather than a 'one-size-fits-all' model. We also need to move away from bordering in both senses of the term: the bordering practices of securitization and violence and the epistemic bordering that enables the exclusion of certain people and non-human animals. We need to acknowledge that the boundaries of species are leaky. We live among other creatures, and zoonoses are illustrations of biomotility – *their* pathogens make *us* up, in the flesh.

Until concerted action to reorder our relations with other living beings and things is taken seriously, the politics of attempting to border the leaky boundaries of species will ensure that pandemics are here to stay (such as swine flu) or that we will face new and/or potentially more deadly viruses. These will be carried in by exploited and increasingly pressured populations of bats, birds, boar, camels, monkeys, mosquitos and possums and nurtured by human poverty and the impacts of climate change (BBC Future, 2021). A sociology of the zoonotic pandemic must insist on foregrounding human social relations with other creatures and living things while also understanding the intra-human hierarchies and their imperatives in social systems that are drivers for increased precarity and vulnerability.

References

Aguirre, A. (2017). Changing patterns of emerging zoonotic diseases in wildlife, domestic animals, and humans linked to biodiversity loss and globalization. *ILAR Journal*, 58(3), 315–18.

Andersen, I. (2020). It is the time for nature: World Environment Day 2020. www.unenvironment.org/news-and-stories/speech/it-time-nature-world-environment-day-2020

Anzaldúa, G.E. (1999). *Borderlands/La Frontera: The New Mestiza*. San Francisco: Aunt Lute Books.

Anzaldúa, G.E. (2015). *Light in the Dark/Luz en lo Oscuro: Rewriting Identity, Spirituality, Reality* (ed. A. Keating). Durham, NC: Duke University Press.

Arcari, P. (2020, 11 April). COVID-19 shows why we need to 'cease and desist' from commodifying animals. *Age of Awareness*. https://medium.com/age-of-awareness/covid-19-shows-why-we-need-to-cease-and-desist-from-commodifying-animals-c042e6eb5be5

Bar-On, Y.M., Phillips, R. and Milo, R. (2018). The biomass distribution on Earth. *PNAS*, 115(25), 6506–11.

Battersea (2020, November). The impact of COVID-19 on companion animal welfare in the UK. https://bdch.org.uk/files/BATTERSEA-Covid-Research-Report.pdf

BBC (2020a, 28 August). Coronavirus: Campaign to encourage workers back to offices. www.bbc.co.uk/news/uk-53942542

BBC (2020b, 31 October). COVID-19: PM announces four-week England lockdown. www.bbc.co.uk/news/uk-54763956

BBC Future. (2021). Stopping the next one: What could the next pandemic be? www.bbc.com/future/article/20210111-what-could-the-next-pandemic-be

Braidotti, R. (2013). *The Posthuman*. Cambridge: Polity Press.

Cassidy, K., Yuval-Davis, N. and Wemyss, G. (2018). *Bordering*. London: Wiley.

Charles, N. (2020, 19 June). The challenges posed by the COVID-19 crisis for human–animal relations. *Toxic News*. https://toxicnews.org/2020/06/19/the-challenges-posed-by-the-covid-19-crisis-for-human-animal-relations

Cudworth, E. (2015). Killing animals: Sociology, species relations and institutionalized violence. *The Sociological Review*, 63(1). https://doi.org/10.1111/1467-954X.12222

Cudworth, E. (2011) *Social Lives with Other Animals: Tales of Sex, Death and Love*. Basingstoke: Palgrave.

Cudworth, E. and Hobden, S. (2011) *Posthuman International Relations: Complexity Ecologism and Global Politics*. London: Zed,

Cudworth, E. and Hobden, S. (2018) *The Emancipatory Project of Posthumanism*. London: Routledge.

Daszak, P. (2020, 9 June). Ignore the conspiracy theorists: scientists know COVID-19 wasn't created in a lab. *The Guardian*. www.theguardian.com/commentisfree/2020/jun/09/conspiracies-covid-19-lab-false-pandemic

DEFRA. (Department for Environment, Food and Rural Affairs) (2020, 27 July). COVID-19 confirmed in pet cat in the UK. *Media release*. www.gov.uk/government/news/covid-19-confirmed-in-pet-cat-in-the-uk

Donaldson, S. and Kymlicka, W. (2011). *Zoopolis: A Political Theory of Animal Rights*. Oxford: Oxford University Press.

Ellis, R.D., McWhorter, T.J. and Maron M. (2012). Integrating landscape ecology and conservation physiology. *Landscape Ecology*, 27, 1–12.

Farmer, P. (2001). *Inequalities and Infections: The Modern Plagues*. Berkeley, CA: University of California Press.

Fine Maron, D. (2020, 15 April). 'Wet markets' likely launched the coronavirus. Here's what you need to know. *National Geographic*. www.nationalgeographic.com/animals/article/coronavirus-linked-to-chinese-wet-markets

Foucault, M. (1991). Governmentality. In G. Burchill, C. Gordon and P. Milet (eds), *The Foucault Effect*. London: Harvester Wheatsheaf.

Friant, S., Paige, S.B. and Goldberg, T.L. (2015). Drivers of bushmeat hunting and perceptions of zoonoses in Nigerian hunting communities. *PLOS Neglected Tropical Diseases*, 9(5), e0003792.

The Guardian. (2020, 19 May). Millions of US farm animals to be culled by suffocation, drowning and shooting. *The Guardian*. www.theguardian.com/environment/2020/may/19/millions-of-us-farm-animals-to-be-culled-by-suffocation-drowning-and-shooting-coronavirus

Hinchliffe, S. (2015). More than one world, more than one health: Reconfiguring inter-species health. *Social Science and Medicine*, 129, 28–35.

Haraway, D. (2016) *Staying with the Trouble: Making Kin in the Chthulucene*. Durham, NC: Duke University Press.

Harvey, J. (2020, 3 June). Jane Goodall: Humanity is finished if it fails to adapt after COVID-19. *The Guardian*. www.theguardian.com/science/2020/jun/03/jane-goodall-humanity-is-finished-if-it-fails-to-adapt-after-covid-19

Hutton, G. (2020, 22 December). *Eat Out to Help Out Scheme*. House of Commons Briefing Paper Number CBP 8978. London: House of Commons Library.

Irvine, L. (2006). Animals in disasters: Issues for animal liberation activism and policy. *Animal Liberation Philosophy and Policy Journal*, 4(1), 1–16.

IPBES (Intergovernmental Science-Policy Platform on Biodiversity and Ecosystem Services) (2020, 25 September). Nature's dangerous decline is unprecedented, but it is not too late to act. https://ipbes.net/news/special-report

Iossifova, D. (2020). Reading borders in the everyday: Bordering-as-practice. In J.W. Scott (ed.), *A Research Agenda for Border Studies*. Cheltenham: Edward Elgar.

Ji, J.S. (2020). Origins of MERS-CoV, and lessons for 2019-nCoV. *The Lancet Planetary Health*, 4(3). www.thelancet.com/journals/lanplh/article/PIIS2542-5196(20)30032-2/fulltext

Jones, B.A., Grace, D., Kock, R., Alonso, S., Rushton, K., Said, M.Y., McKeever, D., Mutua, F., Young, J., McDermott, J. and Pfeiffer, D.U. (2013). Zoonosis emergence linked to agricultural intensification and environmental change. *PNAS*, 110(21), 8399–404.

Jones, K., Patel, N., Levy, M., et al. (2008). Global trends in emerging infectious diseases. *Nature*, 451, 990–93.

Jordan, L. and Howard, E. (2020). *Breaking Down the Amazon: How Deforestation Could Drive the Next Pandemic*. Unearthed. https://unearthed.greenpeace.org/2020/04/24/deforestation-amazon-next-pandemic-covid-coronavirus

Jørgensen, D. (2015). Rethinking Rewilding. *Geoforum*, 65, 482–88.

Keck, F. and Lynteris, C. (2018). Zoonosis: Prospects and challenges for medical anthropology. *Medicine Anthropology Theory*, 5(3). https://doi.org/10.17157/mat.5.3.372

Kim, A. (2020, 15 March). Cats and dogs abandoned at the start of the coronavirus outbreak are now starving or being killed. *CNN*. https://edition.cnn.com/2020/03/15/asia/coronavirus-animals-pets-trnd/index.html

Klein, N. (2007). *The Shock Doctrine: The Rise of Disaster Capitalism*. New York: Knopf.

Klein, N. (2017). *No is Not Enough: Defeating the New Shock Politics*. London: Allen Lane.

Kolbert, E. (2014). *The Sixth Extinction: An Unnatural History*. London: Bloomsbury.

Kolossov, V. and Scott, J. (2013). Selected conceptual issues in border studies. *Revue belge de géographie*, 1, 2–16.

Kushner, J. (2021, 26 January). Why camels are worrying coronavirus hunters. BBC Future – Stopping the Next One: Infectious Disease. www.bbc.com/future/article/20210122-the-coronavirus-10-times-more-deadly-than-covid

Livermore, S. (2020). Ocean noise quiets during the COVID-19 pandemic. www.ifaw.org/people/opinions/ocean-noise-quiets-covid19-pandemic

Łopucki, R., Kitowski, I., Perlińska-Teresiak, M. and Klich, D. (2021). How is wildlife affected by the COVID-19 pandemic? Lockdown effect on the road mortality of hedgehogs. *Animals*, 11(3), 868.

Malm, A. (2020) *Corona, Climate, Chronic Emergency: War Communism in the Twenty-First Century*. London: Verso.

Mezzadra, S. and Neilson, B. (2012). Between inclusion and exclusion: On the topology of global space and borders. *Theory, Culture and Society*, 29(4–5), 58–75.

Middleton, J., Reinjes, R. and Lopez, H. (2020). Meat plants – A new front line in the COVID-19 pandemic. *British Medical Journal*, 370, m2716. https://doi.org/10.1136/bmj.m271

Mignolo, W.D. and Tlostanova, M.V. (2006). Theorizing from the borders: Shifting to geo- and body politics of knowledge. *European Journal of Social Theory*, 9(2), 205–21.

Murray, A. (2020, 11 November). Coronavirus: Denmark shaken by cull of millions of mink. www.bbc.co.uk/news/world-europe-54890229

Nail, T. (2016). *Theory of the Border*. Oxford: Oxford University Press.

Nash, C. and Reid, B. (2010). Border crossing: New approaches to the Irish border. *Irish Studies Review*, 18(3), 265–84.

Nyamnjoh, F.B. (2020). *COVID-19 and the Resilience of Systemic Suppression, Oppression and Repression*. Working paper, summary in F.B. Nyamnjoh, COVID-19: The humbling and humbled virus. Corona Times, 20 April 2020. www.coronatimes.net/covid-19-humbling-humbled-virus

Oberlander, J. (2020). Introduction to COVID-19: Politics, inequalities and pandemic. *Journal of Health Politics, Policy and Law*, 45(6), 905–6.

Parry, J., Young, Z., Bevan, S., Veliziotis, M., Baruch, Y., Beigi, M., Bajorek, Z., Salter, E. and Tochia, C. (2021). Working from home under COVID-19 lockdown:

Transitions and tensions – work after lockdown. *Institute for Employment Studies.* www.employment-studies.co.uk/resource/working-home-under-covid-19-lockdown

Poore, J. and Nemek, T. (2018). Reducing food's environmental impacts through producers and consumers. *Nature*, 360(6392), 987–92.

Prior, J. and Ward, K.J., 2016. Rethinking rewilding: A response to Jørgensen. *Geoforum*, 69, 132–135.

Quammen, D. (2012). *Spillover: Animal Infections and the Next Human Pandemic.* London: Bodley Head.

Quammen, D. (2020, 28 January). We made the coronavirus epidemic. *New York Times.* www.nytimes.com/2020/01/28/opinion/coronavirus-china.html?smtyp=curandsmid=tw-nytopinion

Razai, M., Kanakam, S., Hadyn, K.N., Azeem, M., Aneez, E. and Williams, D.R. (2021). Mitigating ethnic disparities in COVID-19 and beyond. *British Medical Journal*, 372, m4921. https://doi.org/10.1136/bmj.m4921

Rohr, J.R., Barrett, C.B., Civitello, D.J., Craft, M.E., Delius, B., DeLeo, G.A., Hudson, P.J., Jouanard, N., Nguyen, K.H., Ostfeld, R.S., Remais, J.V., Riveau, G., Sokolow, S.H. and Tilman, D. (2019). Emerging human infectious diseases and the links to global food production. *Nature Sustainability*, 2, 445–56.

Romano, J.L. (2020). Politics of prevention: Reflections on the COVID-19 pandemic. *Journal of Prevention and Public Health Promotion*, 1(1), 34–57.

Sassen, S. (2013). When territory deborders territoriality. *Territory, Politics, Governance*, 1(1), 21–45.

Seltmann, A., Gábor Á., Czirják, A. C., Bernard, H., Struebig, M.J. and Voigt, C.C. (2017). Habitat disturbance results in chronic stress and impaired health status in forest-dwelling paleotropical bats. *Conservation Physiology*, 5(1). https://doi.org/10.1093/conphys/cox020

Sit, T.H.C., Brackman, C.J., Ip, S.M. et al. (2020). Infection of dogs with SARS-CoV-2. *Nature*, 586, 776–78.

Shukin, N. (2009). *Animal Capital: Rendering Life in Biopolitical Times.* Minneapolis, MN: University of Minnesota Press.

Solis, M. (2020, 13 March). Coronavirus is the perfect disaster for 'disaster capitalism'. www.vice.com/en/article/5dmqyk/naomi-klein-interview-on-coronavirus-and-disaster-capitalism-shock-doctrine

Stiglitz, J. (2020). The pandemic has laid bare deep divisions, but it's not too late to change course. *International Monetary Fund: Finance and Development.* www.imf.org/external/pubs/ft/fandd/2020/09/COVID19-and-global-inequality-joseph-stiglitz.htm

V (formerly Eve Ensler). (2021, 1 June). How male power fought back. *The Guardian, Life and Arts.*

Tollefson, J. (2021, 15 January). COVID curbed carbon emissions in 2020 – but not by much. *Nature.* www.nature.com/articles/d41586-021-00090-3

van Dooren, T. (2020, 22 March). Pangolins and pandemics: The real source of this crisis is human, not animal. *NewMatilda.* https://newmatilda.com/2020/03/22/pangolins-and-pandemics-the-real-source-of-this-crisis-is-human-not-animal/?utm_campaign=shareaholicandutm_medium=email_thisandutm_source=email

Ward, K.J. (2019). For wilderness or wildness? Decolonising rewilding. In N. Pettorelli, S.M. Durant & J.T. Toit (eds.) *Rewilding.* Cambridge: Cambridge University Press, 34–54.

Westcott, B. and Wang, S. (2020, 23 April). China's wet markets are not what some people think they are. *CNN.* https://edition.cnn.com/2020/04/14/asia/china-wet-market-coronavirus-intl-hnk/index.html

Woods, M. (2005). Ecological restoration and the renewal of wildness and freedom. In Heyd, T. (Ed.), *Recognizing the Autonomy of Nature: Theory and Practice.* New York: Columbia University Press, 170–188.

World Health Organization (2010). The FAO-OIE-WHO collaboration: Sharing responsibilities and coordinating global activities to address health risks at the animal–human ecosystems interfaces: A tripartite concept note. www.fao.org/3/ak736e/ak736e00.pdf

Wolfe, N., Daszak, P., Kilpatrick, A.M. and Burke D.S. (2005). Bushmeat hunting, deforestation, and prediction of zoonotic disease emergence. *Emerging Infectious Diseases*, 11(12), 1822–27.

World Wildlife Fund. (2021). Living planet index. https://livingplanetindex.org/home/index

Yuval-Davis, N. and Stoetzler, M. (2002). 'Imagined boundaries and borders: A gendered gaze.' *European Journal of Women's Studies*, 9(3), 329–44.

Xiao, X., Newman, C., Buesching, C.D., et al. (2021). Animal sales from Wuhan Wet Markets immediately prior to the COVID-19 pandemic. *Scientific Reports*, 11, Article 11898.

9 That killing joke isn't funny anymore

Rebranding speciesism after Brexit

Matthew Cole

Introduction

The sociological analysis of the discursive construction of non-human animals is crucial to understanding the normalization of their oppression (Cole and Stewart, 2016). In previous research, I and co-authors have documented how discourse is symbiotically related to practice, with the language and imagery through which non-human animals are constructed bound up with the ways in which humans behave towards them (see for instance Cole and Stewart 2018; Morgan and Cole, 2011; Stewart and Cole, 2009). These symbiotic relations can be 'mapped' across a spatio-discursive terrain in which most non-human animals are positioned and repositioned by humans, both literally and figuratively, according to their constructed utility/utilities (see Figure 9.1).

A simplified version of Figure 9.1 captures the overall character of the four 'zones' of the map, reproduced in Figure 9.2.

Figures 9.1 and 9.2 are underpinned by Michel Foucault's (1976, 1998) conceptualization of discourse, which may be defined as 'an authoritative system of communication, encompassing language, images and symbols, about a specific subject' (Cole, 2020, p. 137). In the contemporary United Kingdom (and much of the world), authoritative communication about non-human animals tends to be speciesist. That is, human interests are asserted as paramount and the interests of non-human animals are subordinate, or completely elided. This is most obvious in the case of 'food animals', whose fundamental interest in remaining alive is subordinated to the human interest in consuming their bodies or reproductive by-products (see Cole and Stewart, 2016 for a fuller discussion of the linguistic categorization of non-human animals). So, to describe non-human animals as 'discursively constructed' points to the crucial role of discursive authorities in shaping knowledge about non-human animals. A discursive authority is an entity that makes successful truth claims regarding the specific subject(s) about which they communicate. For instance, the 'livestock' industry may be interpreted as a discursive authority about the 'truth' of cows, pigs, chickens and so on as 'food animals'. Discursive authorities are resisted, however, such that their truth claims are contested by competing discourses. In this example, vegan activists and scholars are key sources of

DOI: 10.4324/9781003257912-13

SENSIBILITY

Human
animals

Representations
of animals:
'characters'

'Pets'

Dead
'Meat'

'Entertainment'
Animals

'Working'
animals

SUBJECTIFICATION

OBJECTIFICATION

'Wild'
carnivores

'Laboratory'
animals

'Wild' non-
carnivores

'Farmed'
animals

'Vermin'

NON-SENSIBILITY

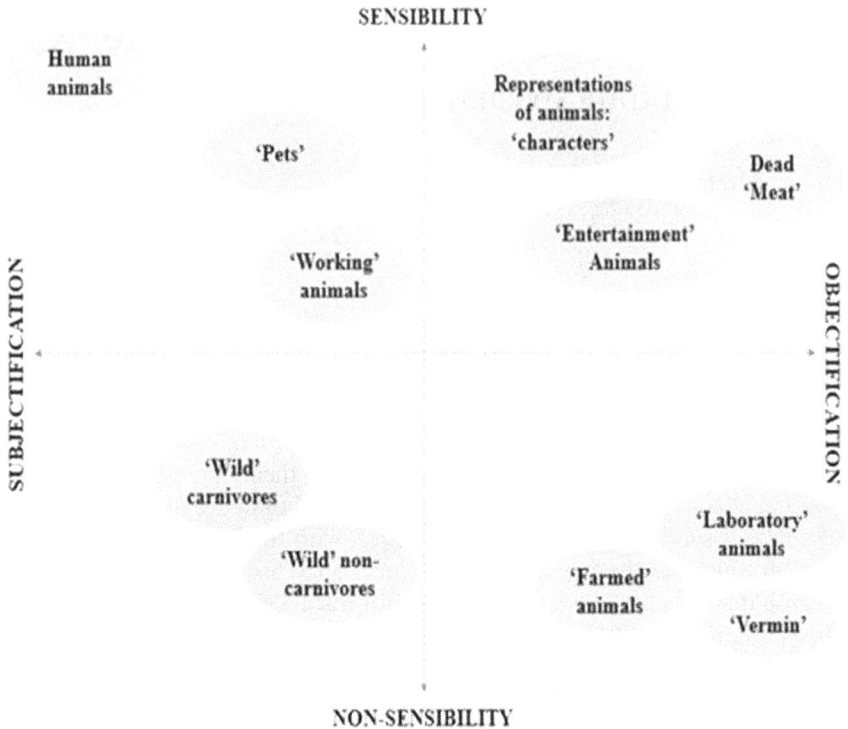

Figure 9.1 Conceptual map of the discursive and physical positioning of non-human
animals.

discursive contestation of the 'livestock' industries' claims, asserting that 'livestock' are really exploited victims of human violence, for instance, rather than
being 'naturally suited' or destined to being food for humans.

In this light, this chapter utilizes the conceptual maps in Figures 9.1 and 9.2
as analytical tools to facilitate a parallel deconstruction of two disparate, yet
cognate discursive constructions to expose the violent absurdity involved in positioning non-human animals in relation to their utility: a recent media campaign
to rename two aquatic species by Cornish 'fishermen' (BBC News, 2021) and the
'laughing fish' storyline from *Detective Comics* (Englehart, 1978a, 1978b) and its
adaptation for television in *Batman: The Animated Series* ('The Laughing
Fish', 1992).

The Cornish Fish Producers Organisation (CFPO) is seeking to expand the
British market for spider crabs and megrim soles by renaming them, respectively, as 'Cornish King crab' and 'Cornish sole'. This is in response to lost
export sales of both species to Europe in the aftermath of Brexit (BBC News,
2021). In 'The Laughing Fish', Batman's arch nemesis, The Joker, brands fishes
with his own facial likeness (achieved through the administration of a chemical
toxin to Gotham City's offshore waters) in order to copyright them and thereby

SENSIBILITY

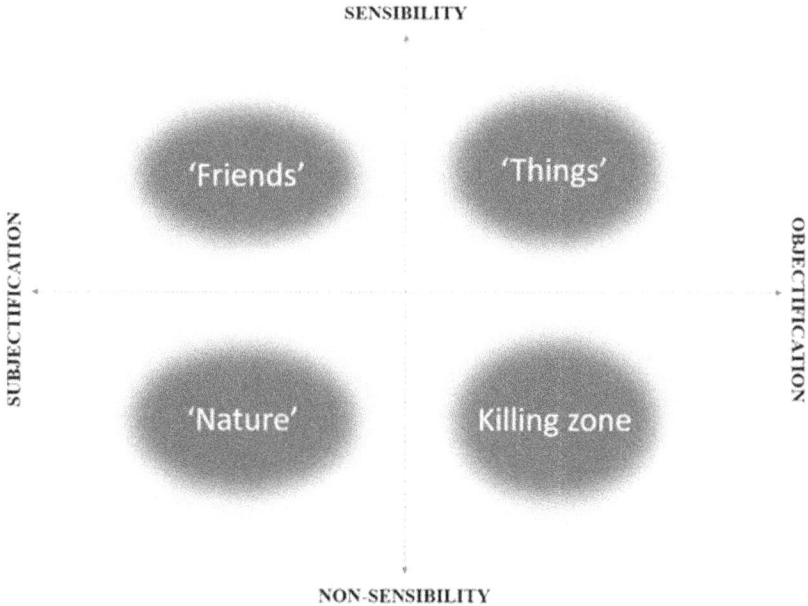

SUBJECTIFICATION

OBJECTIFICATION

'Friends'

'Things'

'Nature'

Killing zone

NON-SENSIBILITY

Figure 9.2 Simplified conceptual map of the discursive and physical positioning of non-human animals.

profit from their sale. In both cultural artefacts, human interests are paramount and the interests of the victimized species are erased. Both examples share further common features: they highlight a cultural ambivalence towards the human consumption of at least some aquatic creatures, related to apparent feelings of revulsion aroused by their physiological otherness. Both also therefore dramatize a source of disruption to the acceptability of continued exploitation. In other words, the visual strangeness of fictional laughing fishes, and real spider crabs or megrim soles, are prima facie motivations to resist their categorization as 'food'. The trajectory of the drama works in opposite directions, however; the laughing fishes are estranged by the intervention of The Joker, whereas spider crabs and megrim soles are de-estranged by the intervention of the CFPO. Their 'Cornish' renaming makes them more sensible (to potential British consumers) as cohabitants of a post-Brexit Britain, but simultaneously objectifies them more intensely through the discarding of their pre-Brexit 'megrim' or 'spider' identities that kept them relatively safe from consumption by those same British consumers.

Despite these opposite trajectories, both discourses resolve their respective dramas by reframing aquatic species as legitimate targets of exploitation – in other words, neither is ultimately interested in troubling the speciesist social order. However, that potential remains latent within the fictional and real examples (as it does in all the examples analysed in previous research). Although it is unlikely that any potential consumer of megrim soles or spider crabs is reminded of the comic book storyline, one argument being made here is that speciesist popular culture can predispose consumers to unquestioningly accept the ontological transformation of species, even when the epistemological fraud that underpins it is played out in front of their eyes. In fact, the rebranding of spider crabs and megrim soles enrolled consumers as research participants to inform the rebranding (BBC News, 2021). One purpose of this chapter is to reassert the value of highlighting the absurdity of this process to leverage public awareness and reflexive critique of the uncritical acceptance of speciesist discourses and stimulate both anti-speciesist individual change and support for anti-speciesist structural change.

In order to pursue these aims, the chapter proceeds with a sequential analysis of the CFPO campaign in relation to Figures 9.1 and 9.2, followed by a similar analysis of the 'laughing fish' storyline, in which parallels will be drawn between the two examples.

The discursive beautification of victims of human violence

Using Figures 9.1 and 9.2 as analytical tools, the discursive constructions of 'spider crab' and 'megrim sole' are apparent barriers to repositioning these creatures from the south-west (as inhabitants of 'nature' outside human domination) and into the north-eastern (as objectified 'products') and south-eastern (as victims of lethal violence) terrain, from the perspective of potential British consumers. One of the key processes involved in the positioning of non-human animals in Figures 9.1 and 9.2 is the perverse deployment or inculcation of affectivity, such that 'elevating' representations of living non-humans to the north-eastern quadrant facilitates the murderous repositioning of real non-humans from the south-western 'nature' zone to the south-eastern 'killing' zone (Stewart and Cole, 2009). In simple terms, encouraging people to feel affection towards discursive representations is profitable because it leads to a greater willingness to buy animal products associated with the representation. The negative connotations of 'spider' and the 'grim' syllable in the soles' name inhibit the inculcation of affect towards them. This process is laid bare in a BBC News website report on the CFPO rebranding campaign, which ran with the headline 'Brexit: "Under-loved" Fish Renamed for British Tastes' (BBC News, 2021). The quoted portion of the headline came from a statement by Paul Trebilcock of the CFPO: 'The two species are particularly under-loved in this country but really popular with some of our export markets' (cited in BBC News, 2021). Elsewhere, Trebilcock is reported as claiming that, 'There's this negative thing with megrim – it's a "grim" connotation' (*The Guardian*, 2021). Both examples, from the BBC and *The Guardian*, are indicative of the extent to which news media content can be shaped by 'pre-packaged news', such as those provided by public relations' press releases (Lewis et al., 2008, p. 2). The ensuing analysis suggests that, in this case, news media failed to provide or platform counter-discourses, instead amplifying the CFPOs claim to discursive authority.

In the BBC report, Trebilcock is also reported as asserting that, 'There is something about the names that has negative connotations'. In the 'under-loved' statement, object and subject are confused and conflated – Trebilcock, of course, means 'love' in the sense of gustatory enjoyment of the consumption of the flesh of these creatures not 'love' in the sense of positive affect for the living animals. The 'negative connotations' of the species' names are negative from the perspective of the industry that seeks to exploit them; conversely, the names are *protective* of these species insofar as they dissuade consumers from accepting their discursive positioning as 'food'. Human 'love' in the bastard-ized sense that Trebilcock invokes is a very dangerous thing for non-human animals. If the living creatures were genuinely loved – as *someones* rather than *somethings*, to paraphrase Carol J. Adams (2015) – they would be discursively unkillable. Trebilcock's statement therefore reveals the casual speciesism and Orwellian doublethink involved in overtly asserting 'love' for living creatures as an easily understood metaphor for 'loving' to eat their flesh – easily understood because of the broader speciesist context in which violence towards non-human animals is normalized. The latter could predispose people to care for and protect these creatures, rather than consume them – a risky outcome for the CFPO. This can be highlighted by invoking an old joke and substituting in the CFPOs target species:

Original joke: 'I love children, but I couldn't eat a whole one.'
New 'joke' 1: 'I love Cornish soles, but I couldn't eat a whole one.'
New 'joke' 2: 'I love Cornish king crabs, but I couldn't eat a whole one.'

In the original joke, the comic effect emerges from the discursive unconsum-ability of (implicitly human) children. In the new versions, there is no comic effect in a speciesist context within which 'Cornish soles' or 'Cornish king crabs' are discursively consumable. In fact, the latter 'jokes' read more as state-ments of fact: that the speaker is physically incapable of consuming that quan-tity of soles' or crabs' flesh. If we inhabited a vegan, anti-speciesist culture, the latter jokes might actually function as jokes, because the notion of eating soles or crabs of any name would be self-evidently ridiculous. Alas, the killing jokes aren't funny (yet).

This begs the question of exactly how the CFPO's intended victims are dis-cursively repositioned in the north-eastern zone from the perspective of a puta-tive British consumer. To start with spider crabs, they spatially occupy the 'nature' zone of the conceptual map – that is, they live freely in an aquatic environment, outside human control, though not free from the harmful envi-ronmental impacts of destructive human activity. This spatial location can be represented in the style of photography featured in Figure 9.3.

As with many 'wild' animals commercialized as products, the living creature in Figure 9.3 is both spatially and discursively very distant from the plate of a human consumer. As a self-evidently living creature in a natural environment, this crab is represented with some degree of subjectivity, well suited to flourish

Figure 9.3 A spider crab or a Cornish king crab?

in this kind of environment. However, such representations are dangerous and can still function as a means of rendering non-human animals as consumable spectacles – objects of curiosity rather than agential subjects. Therein lies a clear route to the commodification of non-human animals from the 'nature' zone as spectacles in 'zoos', 'sports', 'wildlife' tourist experiences and so on, wherein their intrinsic value as subjects of life is subordinated to their trivial use value to humans in the north-eastern zone of Figures 9.1 and 9.2. The curiosity 'value' of spider crabs is highlighted in a *Guardian* newspaper article about the CFPO campaign. Written as a comedic dialogue between an imagined ignorant interlocutor and the authoritative voice of *The Guardian*, the interlocutor in the article describes spider crabs as resembling 'massive armoured sea spiders' (*The Guardian*, 2021), but that their 'Arachnophobia [is] gone' in response to the rebranding of spider crabs as 'Cornish king crabs'. It is this renaming that effects the shift from the south-west to the south-eastern zones for the real animals, under the discursive cover of the novel appellation being inserted into the north-eastern zone. This process is represented in Figure 9.4.

This is a recent example of a very common process, by which fictional representations of non-human animals act as magnets for human affectivity at the same time as they attenuate the sensibility of the real animals they represent, and by so doing legitimate the violence inherent in transforming real living animals into consumable products (Stewart and Cole, 2009). Visual culture is especially effective here, as in the innumerable instances of 'suicide food' (see Cole, 2011) – cartoonified images of anthropomorphized non-human animals who desire, or even participate in, the consumption of their own flesh (such as cartoon pigs advertising 'pork' sausages). What the spider crab and megrim

SENSIBILITY

The brand:
Cornish soles
and king crabs

Commodified
flesh

'Friends'

'Things'

SUBJECTIFICATION

OBJECTIFICATION

'Nature'

Killing zone

Megrim
soles and
spider
crabs

Victims
regardless
of name

NON-SENSIBILITY

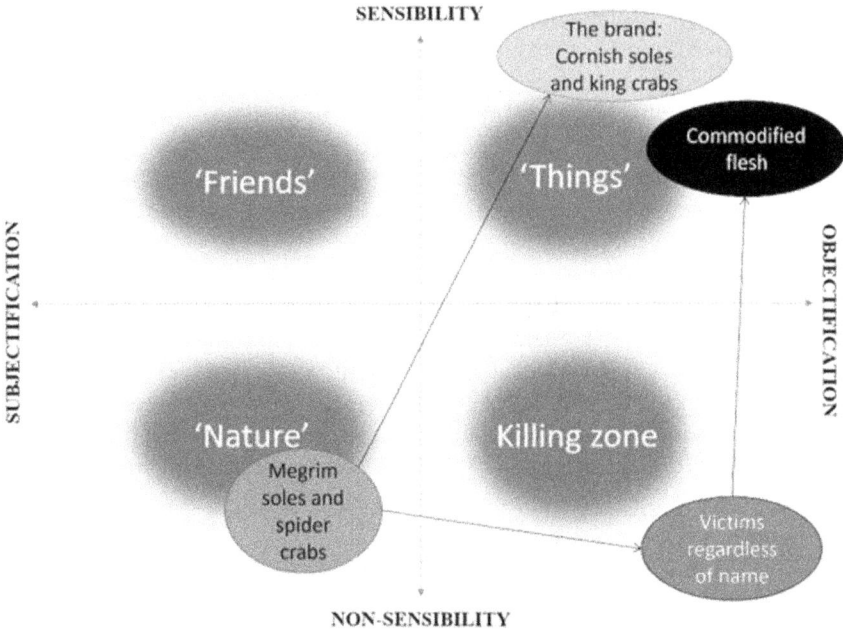

Figure 9.4 Repositioning process.

sole examples demonstrate is that visual representations are not always necessary to effect the repositioning of non-human animals in the conceptual map; linguistic manipulation can be powerful enough. This may be related to the greater degree of physiognomic otherness of these species. That is, the fact that their anatomy appears different from human anatomy to a more significant degree, compared with mammals such as pigs or cows, means that less discursive work is required to render them killable and consumable. Simply put, a name change will suffice without producing the types of cutified marketing images that distract from the fates of real 'livestock' (Stewart and Cole, 2011).

The Guardian article lays bare the absurdity of renaming these species in order to make them newly palatable. It does this partly by invoking historically lethal/profitable rebranding campaigns that renamed 'pilchards' as 'Cornish sardines' and 'bull huss' as 'rock salmon'. In respect of the current campaign, the article's headline refers to megrim soles as 'this ugly fish', and early in the piece the interlocutor states, 'So it's a fish. Looks horrid', which receives the reply, 'Not at all; they love it in Spain.' This brief reference to the Spanish market for the product reveals the contingency of the positioning of non-human animals in Figures 9.1 and 9.2 (Cole and Stewart, 2016); the intrinsic characteristics, preferences and interests of the species subject to positioning and repositioning are largely irrelevant – it is human meaning-making practices that authorize the (in)utility of other species in specific spatial and temporal contexts. So, *The Guardian* itself (as well as the BBC and other media outlets that reported the story) goes beyond reporting on the rebranding campaign and becomes complicit in the marketing ploy. In the case of *The Guardian*, the authoritative voice declaiming that the 'Cornish sole' is, 'Quite good, actually – a quality, sustainable white fish' and that it's a 'Good choice' on the part of the interlocutor to anticipate that, 'I'll have a dressed one ['Cornish king crab'] for starters, followed by the grilled Cornish sole with lemon butter.' Similarly, but without the 'comedic' overtones, the BBC enjoins an authoritative voice to extol the virtues of the renamed species:

> Chef Jack Stein … said spider crab and megrim sole were both 'good value and delicious'. He said spider crab was very sustainable with 'amazing quality meat and there is just so much of it in the summer' but lots of it was sold in the EU. Similarly megrim sole had a 'lovely taste' similar to plaice, he said. 'With a bit of a rebrand and a bit of a publicity drive we might be able to eat more of it in this country,' he added.
>
> (BBC News, 2021)

Neither article gave a voice to advocates for spider crabs or megrim soles, whether in terms of welfare issues raised by their victimization for food, the ethical issues involved in their denial of a right to life or vegan critiques and alternatives to consuming them at all. Stein's contribution is typical of the deindividuating massification of commodified non-human animals, focusing on the 'sustainability' of the species as a whole without concern for the sustaining of individual crabs' lives. This is compounded with the reference to crabs' flesh as 'meat' being abundantly available, eliding the violent human activity that underpins that abundance. Furthermore, Stein refers to the consumption of flesh ('it') and not the consumption of killed creatures (which would necessitate a reference to being 'able to eat more of *them*'). Once again, someones are discursively constructed as objectified somethings. As journalism, these examples illustrate the biases that predominate in mainstream reporting of the exploitation of non-human animals (Freeman, 2016), as well as supporting Lewis et al.'s findings of the prevalence of public relations agenda-setting in UK news media (2008, pp. 8–10). Moreover, they illustrate the close connection between the media and industry sectors of the animal–industrial complex (Noske, 1999). This association between meaning-marking (representing non-human animals) and social practices (killing them for profit) is another way of highlighting what Figures 9.1 and 9.2 are designed to capture: the indissociable symbiosis of representations and practices in the reproduction of the speciesist social order. However, a final stage in the repositioning of megrim soles and spider crabs remains to be examined: their discursive appropriation as regionalist/nationalist symbols in the context of Brexit.

In both cases, the negative connotations of 'megrim' or 'spider' (connotations that I have argued above are actually 'positive' insofar as they constructed these species as less killable) are replaced with 'Cornish', with the crabs 'benefiting' from the further appellation of 'king'. The latter implicitly assigns a masculine gender to the entire species, reproducing a trope of associating carnivory with hegemonic masculinity (Adams, 2015; Cudworth, 2011). In other words, the kingly status of the crabs augments the desirability of their flesh by invoking the magico-cultural norm that flesh consumption confers the (constructed) status of the victim on the consumer (Fiddes, 1991). 'Cornish' alludes to Cornwall, the most south-westerly county of England, renowned for its natural coastal beauty and as a tourist destination that provides a picturesque escape from urban life. This reputation is ironic given the representation of 'nature' as the south-western zone of Figure 9.2. However, the context of Brexit adds a further layer of meaning. As the BBC News (2021) article points out, 'Each year about 1,000 tonnes of megrim sole is brought into Newlyn, Cornwall's biggest fish market, with 98% exported. About 85% of spider crab is exported, mainly to Spain'. The deindividuating measurement of aquatic victims by weight rather than as individual creatures is noteworthy in itself, but the article reports these figures in the context of new export difficulties resulting from Brexit, necessitating the cultivation of new domestic markets to sustain this sector of the Cornish fishing industry. In that light, the 'Cornish' appellation implicitly enjoins consumers to eat these creatures as a patriotic duty: 'Mr Trebilcock said: "Our investigation revealed that simply by calling it Cornish sole, straight away more people were willing to try it and were more interested in finding out where it came from"' (BBC News, 2021). The soles are de-gendered and thereby deindividuated once again ('it' not 'them' or 'they'), but redefined in relation to a politicized human regional/national context. Discourse is manipulated to ensure their continued victimization despite the politically created market barrier. It will not matter at all to the crabs and soles whether they are ultimately eaten by British or Spanish consumers, but what matters to the industry is that they will be eaten by someone. What this example also demonstrates is the assumed right of the CFPO to rename these species, an illustration of the more generalized and unquestioned speciesist assertion of human interests over those of other species. As noted above, media reporting colluded in the CFPO's assertion, supporting its interests to the complete exclusion of those of megrim soles or spider crabs.

   In summary, the CFPO's renaming of these species is absurd yet effective. It brings into stark relief how meaning-making practices construct non-human animals as killable, commodifiable and marketable, sustaining exploitative industries and reproducing the speciesist social order. One of the key insights of Foucault's development of a theory of discourse was that meaning is written on the surface, undisguised and brazen, contrasting with the Marxist assumption of buried ideological underpinnings of cultural meanings (Foucault, 1976). Nowhere is this clearer than in the CFPO's public and deliberately publicized redefinition of its victim species for the explicit purpose of continuing to kill and profit from them. This process is thoroughly normalized – or, to put it another way, speciesist discourse is extremely robust. In order to further emphasize its absurdity, the analysis now turns towards the 'laughing fish' storyline from *Detective Comics*, a narrative presented as self-evidently absurd, but in key respects very similar to the actions of the CFPO and count-less other discursive authorities in the animal–industrial complex.

The laughing fish

The 'laughing fish' storyline appeared across issues 475, entitled 'The Laughing Fish' (Englehart, 1978a) and 476, entitled 'The Sign of The Joker' (Englehart, 1978b) of *Detective Comics*, respectively, published in February and April 1978. *Detective Comics*, from which the DC brand took its name, began publication in 1937, hosting the first appearance of Batman in 1939 and remaining in almost continuous publication since that date to the present, now running to well over 1,000 issues (DC Database, 2021). The storyline was adapted as the 34th produced episode of *Batman: The Animated Series* (BTAS) in 1992 (sharing the title 'The Laughing Fish' with the original comic). BTAS is renowned as a high watermark in animated television, and Batman and his nemesis The Joker continue to enjoy prominence as icons of popular culture, with a new live-action Batman film awaiting imminent cinema release at the time of writing (*The Batman*, 2022), and Joaquin Phoenix having gained critical plaudits for his recent cinematic portrayal of The Joker (*Joker*, 2019). A hallmark of The Joker's character is his hatching of bizarre criminal schemes that have comedic elements as well as harmful, sometimes lethal consequences. These schemes are interpreted as evidence of his irrationality, or 'madness', juxtaposed with Batman's attempt to reassert the rational, lawful order of the fictional Gotham City that provides the backdrop for their conflicts. The laughing fish storyline exemplifies this dynamic. In both the comic and animated versions of the story, The Joker chemically alters fish's faces in order to profit from asserting intellectual property rights over them because they bear his own rictus-grin likeness. Denied such property rights, The Joker murders two officials whom he holds responsible, despite Batman's attempts to protect the threatened bureaucrats. In *Detective Comics* 475, the scheme first comes to light when Batman is confronted by a fisherman at Gotham docks:

Fisherman 1: 'Batman! Our Fish! Our Fish!'
Batman: 'Get a grip on yourself, man! What are you yelling about?'
Fisherman 1: 'Look!' [he holds up a wicker basket filled with fishes with anthropomorphized snarling expressions, white faces, red lips and The Joker's distinctive open-mouthed rictus grin, displaying sharp teeth].
Batman: 'What the …! All of them … with The Joker's face?!?'
Fisherman 1: 'All of 'em! And it's the same for everybody else!' 'Our whole catch is contaminated with that lunatic grin! Herring – cod – ! All of 'em laughing at us!'
Fisherman 2: 'It scares the britches off me!'

Batman:	'It's supposed to! Fear clouds your mind, and that's The Joker's strongest weapon!'
Fisherman 1:	'But what does he want? Why would he make the fish look like him?'
Batman:	'That I don't know! With ordinary men, you might figure some motive ... But The Joker's mind is clouded in madness! His motives make sense to him alone!'

(Englehart, 1978a, p. 6)

This exchange highlights how The Joker's intervention disrupts the speciesist social order insofar as it disturbs this particular operation of the animal–industrial complex. As the first fishermen asserts, these are 'our' fish, extracted from their natural environment, killed and objectified as commodities – the instrumentally rational business as usual of repositioning fishes from the south-western region of Figure 9.2 to the north-east, via the obfuscated killing zone of the south-east (in neither version of the story are fishes shown actually being killed; only alive or already dead; this is in common with imagery of spider crabs and megrim soles used by the *BBC* and *Guardian* articles discussed earlier). The Joker disturbs this process by criminally disputing the property rights of the animal–industrial complex. In the comic version of the storyline, the laughing fish phenomenon encompasses both seaboards of the United States, rather than being a localized offshore Gotham City incident, and in issue 476, The Joker declares his global ambitions in a soliloquy: 'Today the American fish – and tomorrow, all the fish in the world!' (Englehart, 1978b, p. 5). In the animated episode, Gotham's docks include a warehouse adorned with a typical 'suicide food' image: a sign showing an anthropomorphized smiling fish, using a fin to wave at the human spectator of the image, standing upright and advertising 'fresh fish'. This is a 'legitimate' representation, however – a cultural aspect of the animal–industrial complex that populates the north-eastern zone of Figure 9.2, deployed to inculcate affectivity towards the products of violence while distracting from the real victims of that violence in the south-eastern killing zone, as well as suggesting victims' willing complicity in their own deaths. This is commonplace in real-world advertising of animal products (although not the case here, such representations are also often sexualized, as documented by Carol J. Adams (2020) in *The Pornography of Meat*.)

By contrast, The Joker directly alters the victims' physiognomy rather than altering their representation, with their threatening snarl contrasting with the cosy complicity of the painted fish sign. The effect of this on the second fisherman is expressed in his admission of fear: 'It scares the britches off me!' while the first fisherman testifies to the discomfiting experience of being interpellated by fishes themselves: 'All of'em laughing at us!' Although this discomfiture of fishermen is not explored further in the story, the horror of being laughed at hints at a provocation of guilt felt by this fisherman for his violent occupation, the anthropomorphization lending the fishes a post-mortem capacity to haunt him for his role in killing sentient beings. By contrast, real aquatic creatures are given precious few opportunities to confront their human exploiters with expressions of their sentience and suffering, with their experiences of being wrenched from their home invisibilized and silenced beneath the surface of the ocean (the ocean's surface is both a real and metaphorical boundary between the southern and northern regions of Figures 9.1 and 9.2 for aquatic creatures). The horror inspired by the laughing fishes arguably derives from The Joker collapsing the boundaries between the regions of Figure 9.2, hybridizing a self-serving representation (fishes branded with his image) with the deaths of 'real' fishes (real from the perspective of the story's characters that is). The interpellatory gaze of the fishes confronts the fictional fishermen with an undeniable reminder of subjectivity, of victims of violence confronting their killers with their frozen laughter. At the same time, The Joker's poisoning of the fishes to distort their faces is perpetrated without guilt – the villain's purported madness means he has no need of the obfuscatory discursive architecture represented in Figures 9.1 and 9.2. Free of the social norm that requires the maintenance of a positive self-concept of care and concern for non-human animals (Cole and Stewart, 2016), The Joker freely perpetrates and exposes the violence that always underpins the duplicitous use of 'happy' representations of the victims of the animal–industrial complex. However, while it is possible to discern critical anti-speciesist potential in the laughing fish story, this is not The Joker's motive: he seeks to usurp the animal–industrial complex, not to dismantle it.

The Joker's self-interested motive is made clear when he enters the story. In both the comic and animated episode, The Joker confronts a patent clerk (named Mr Francis) with his claim. In the comic, he approaches Francis in his office to 'make arrangements':

Francis: 'Wh-what arrangements?'
Joker: 'For my fish, of course!' [he throws a dead 'laughing' fish onto Francis' desk]
…
Francis: 'Uh … your fish?'
…
Joker: 'Everybody's talking about my Joker-fish! They all recognize the face – it's my fortune, even on a flounder's fizz – and since I plan to continue secretly dumping the chemical [along with the plot exposition!] that gives the fish my face, the little finnies are permanently identified with me! No matter what they once were, they're just Joker-fish now! Sooo … once we fill out all your tedious copyright forms – I'll get a cut of every fish-sale in America! A nickel per fish-sandwich – fifty cents for filet of sole!'
…

Francis: 'Nobody can copyright fish – or even fish faces! They're a natural resource!'

…

Joker: 'But the fish share my unique face! If colonel what's-his-name can have chickens, when they don't even have mustaches – ! And you deny this to me! You see why I am forced to crime! You have until midnight to change your mind, Francis! If you don't, you'll be the poorest fish of all – and dead as a mackerel! Hahahahaha.'

Francis [thinks]:   'He's – He's insane!'

(Englehart, 1978a, pp. 8–9)

In the animated episode, the scene plays out similarly, with The Joker asserting that 'Since every fish in Gotham now bears my famous and frankly fabulous face, I should be getting a profit from every fish product sold. Let's say, a nickel per fish sandwich, 50 cents for sardines.' The astonished Francis replies, 'No one can copyright fish. They're a natural resource', to which The Joker counters, 'But they share my unique face. Colonel what's his name has chickens and they don't even have mustaches' ('The Laughing Fish', 1992). The criminality of The Joker's plot boils down to theft – appropriating a 'natural resource' and thereby depriving the 'legitimate' animal–industrial complex of its assumed right to exploit fish. By invoking the KFC brand, The Joker aligns himself with the clownish emblems of the US fast-food giants (such as Colonel Sanders, Ronald McDonald and Burger King) and highlights the key discursive role of 'benevolent' representations of the masters of the animal–industrial complex. The harm he perpetrates towards the fishes is of no account (no more than is the harm perpetrated by KFC against chickens); whatever suffering may be supposed to result from the alteration of their faces is not mentioned in either the comic or animated versions of the story.

In this respect, the laughing fish storyline adheres to the wider speciesist social order – fishes are utilized as representations of The Joker's villainy and are absent from the narrative as subjects with their own interests and preferences. So, despite his vaunted madness, The Joker's scheme is remarkably conventional in this sense: it depends on the mundanity of the asserted speciesist right to exploit non-human animals in disregard of those animals' own interests. The irrationality of his plot inheres in The Joker's poor grasp of property law, rather than any challenge to speciesism itself – the ghostly subjectivity he lends to the fishes is incidental to his scheme. But irrationality is more clearly evidenced by this incidental subjectivity – that is, The Joker's failure to recognize that the laughing fish are disturbing. It is his assumption that Joker fish would be marketable that is 'mad', because his visage – grotesque as it is – nevertheless humanizes fishes to the extent that they are discursively repositioned, simultaneously monstrous and uncomfortably close to transgressing the north-western zone of Figure 9.2. Their anthropomorphized faces render them unconsumable, hinting at cannibalism and provoking revulsion among potential consumers. In the animated episode, Alfred (Batman's erstwhile guardian and now elderly butler), comments, 'Maybe he's [The Joker] trying to make us all die from disgust?' ('The Laughing Fish', 1992). Recall that the relative 'monstrousness' of 'grim' soles and 'spider' crabs was protective of those species, from potential British consumers. The Joker's plot inverts the beautification of victims achieved by the CFPO and negates the possibility of his own financial success (even if it were legal) by associating fishes with his own monstrous (yet human) personality and physiognomy.

The animated episode plays up the comedic implications of The Joker's lack of awareness of the irrationality of marketing 'Joker-fish' through a television advert (absent from the comic version of the story), witnessed by Batman and Alfred watching from the Batcave (Batman's secret lair). The advert features The Joker alongside his devoted (and abused) girlfriend, Harley Quinn (a young White female character, dressed in a red and black harlequin costume, created for BTAS and absent from the 1978 comic) and two White male henchmen, in a staged kitchen scene. As the advertisement plays, Harley sings a voiceover:

Harley:	'They're finny and funny and oh so delish. They're joyful and folly Joker Fish [packaged anchovies, chopped clams and frozen fish fillets are shown on screen].
Joker:	'Say, Mom, wondering what to feed the family tonight?'
	Harley ('Mom' in the scene) is in a kitchen set with the two henchmen dressed as a boy and girl waiting for dinner.
Harley:	'What'll I feed the family tonight?'
Joker:	'Arr. Try me famous Joker Fish.'
	Joker is wearing oilskins and a fishing-hat, and holds a Joker-fish up to the camera.
Joker:	'There's Smiling Smelt, Giggling Grouper and Happy Haddock.'
	He empties a bag of dead fishes onto a kitchen table.
The watching Alfred interjects:	'This could cause a stampede to pork.'
	The advert continues, with The Joker proffering a forkful of fish to Harley.
Joker:	'Yummy yum-yum.'
	Harley recoils.
Joker [menacing]:	'Eat it.'
	The fish's smiling face is still visible on the plate.
Harley:	'Uh, Mr. J, I have this little problem with the fish.'
	Joker forces a fork into Harley's mouth, and Harley looks to camera.

Harley [unconvincingly]:	'Yummy yum-yum.'
	Harley then exits off camera looking as though she will vomit.
Joker [to camera]:	'Yes, friends, that's Joker fish.'
	Harley retches off camera.
Joker:	'Tasty, tempting and of course naturally low in cholesterol.'
	The camera pans to the henchmen dressed as a boy and girl as the advert ends.

('The Laughing Fish', 1992)

The fictional advertisement can be interpreted as a parodic version of the staged dialogue in *The Guardian* and the enrolment of the authoritative perspective of the chef, Jack Stein, in the BBC story. All three examples attempt the same task: persuading consumers to eat newly rebranded aquatic creatures. However, the CFPO is lauded for its discursive beautification, while The Joker is ridiculed for his discursive (and real) uglification. It is not the exploitation of the fish that is morally or aesthetically ugly, but his tainted brand. This is reinforced by Alfred's interjection: the discomfiture aroused by The Joker's scheme merely diverts the consumer towards another species of victim (euphemistically objectified as 'pork' rather than 'pigs'), rather than away from consuming non-human animals altogether. In the comic version of the story, The Joker has a moment of insight when he realizes this risk: 'But – what if everybody stops eating fish? I hadn't thought of that.' (Englehart, 1978b, p. 5). However, his madness is reasserted when he thinks of an alternative: 'I could use my chemicals on the cattle! Joker-burgers! Outrageous!' (Englehart, 1978b, p. 5). He reaches a similar conclusion in the animated episode: 'You're right Harley, fish are disgusting, I think I'll start using my toxin on cattle. Joker Burgers, ha! Talk about a Happy Meal.' ('The Laughing Fish', 1992). The reference to the McDonald's brand is clear, as is the parallel between The Joker's own costume and makeup and the clown character of Ronald McDonald. The Happy Meal has frequently made use of tie-ins with animated children's films, offering plastic toy versions of animal characters alongside 'animal products', drawing children's affectivity towards the representation and away from the real exploited victims whose remains sit in the same Happy Meal box (Stewart and Cole, 2009). As argued in previous work, the deployment of representations such as these effects a conceptual separation between fictional and real non-human animals, enabling the maintenance of positive self-concepts while vicariously participating in violent exploitation. The image of cows altered so as to bear The Joker's face would collapse that conceptual separation (such 'Joker-cows' are not actually depicted in either iteration of 'The Laughing Fish'), visibilizing cows as victims of violence. The Joker's scheme is 'mad' not only because of its inevitable failure, but more profoundly because it makes it more difficult to maintain a conceptual separation between 'happy' representations and real victims in the animal–industrial complex. Unfortunately, the potential to leverage this implicit critique is not pursued in 'The Laughing Fish', which remains fixed within a speciesist cultural frame.

Conclusion

This analysis demonstrates that widely disparate iterations of speciesist discourse share a common pattern. Across the thousands of miles of the Atlantic Ocean separating the fictional Gotham City from Cornwall, a temporal span of 43 years between the publication of the *Detective Comics* storyline and the CFPO's marketing campaign; and articulating the genres of comic books, television animation, news, public relations and marketing, the exploitation of aquatic species remains remarkably similar. The Joker and the CFPO exemplify the arrogated anthropocentric right to ontologically redefine other species in order to profit from their exploitation. They both stimulate heightened sensibility of 'their' products through engaging the mass media: news outlets report on the CFPO and television news reports on The Joker's scheme in the comic, while The Joker hijacks a television broadcast to show his Joker-fish advertisement. The CFPO and The Joker thus draw attention to the north-east region of Figures 9.1 and 9.2, attempting to articulate their 'products' with positive effect towards their brands. Where the examples differ is in The Joker's failure to successfully maintain the boundaries that demarcate the different terrains of Figures 9.1 and 9.2. Therein lies the comedy and 'madness' of his doomed caper: his defective brand and the collapsing of the conceptual distance between the brand and the real victims of violence. But through this apparent failure, 'The Laughing Fish' suggests a valuable insight precisely because the killing joke is deadly serious for real aquatic victims of human violence: that the business-as-usual operations of the animal–industrial complex and their media portrayal are madder than fiction. We inhabit a culture that remains so thoroughly speciesist that it is not only possible but *normal* to publicize a deliberate ploy to entice consumers to become complicit in new forms of violence, and to feel good about doing so.

References

Adams, C.J. (2015). *The Sexual Politics of Meat – 25th Anniversary Edition: A Feminist-Vegetarian Critical Theory*. New York: Bloomsbury.

Adams, C.J. (2020). *The Pornography of Meat: New and Updated Edition*. New York: Bloomsbury.

The Batman (2022). *Feature film*. Directed by M. Reeves. Burbank, CA: Warner Bros Pictures.

BBC News. (2021). Brexit: 'Under-loved' fish renamed for British tastes. www.bbc.co.uk/news/uk-england-cornwall-55996938

Cole, M. (2011). From 'animal machines' to 'happy meat'? Foucault's ideas of disciplinary and pastoral power applied to 'animal-centred' welfare discourse. *Animals*, 1(1), 83–101.

Cole, M. (2020). Criminology, harm and non-human animals. In L. Copson, E. Dimou, and S. Tombs (eds), *Crime, Harm and the State Book 1*. Milton Keynes: Open University Press.

Cole, M. and Stewart, K. (2016). *Our Children and Other Animals: The Cultural Construction of Human–Animal Relations in Childhood*. London: Routledge.

Cole, M. and Stewart, K. (2018). Socializing superiority: The cultural denaturalization of children's relations with animals. In A. Cutter-McKenzie, K. Malone, and E.B. Hacking (eds), *International Research Handbook on ChildhoodNature: Assemblages of Childhood and Nature Research*. New York: Springer.

Cudworth, E. (2011). *Social Lives with Other Animals: Tales of Sex, Death and Love*. Basingstoke: Palgrave Macmillan.

DC Database. (2021). *Detective Comics* (1937–present). https://dc.fandom.com/wiki/Detective_Comics_Vol_12

Englehart, S. (1978a). The laughing fish. *Detective Comics*, no. 475.

Englehart, S. (1978b). The sign of The Joker. *Detective Comics*, no. 476.

Fiddes, N. (1991). *Meat: A Natural Symbol*. London: Routledge.

Foucault, M. (1976). Truth and power. In M. Foucault, *Power: The Essential Works of Michel Foucault 1954–1984*. Harmondsworth: Penguin.

Foucault, M. (1998). *The Will to Knowledge: The History of Sexuality, Volume 1*. Harmondsworth: Penguin.

Freeman, C.P. (2016). This little piggy went to press: The American news media's construction of animals in agriculture. In N. Almiron, M. Cole and C.P. Freeman (eds), *Critical Animal and Media Studies: Communication for Non-human Animal Advocacy*. New York: Routledge.

The Guardian. (2021, 9 February). Seafood surprise: Could rebranding this ugly fish as 'Cornish sole' make Brits eat it? www.theguardian.com/food/shortcuts/2021/feb/09/seafood-surprise-could-rebranding-ugly-fish-as-cornish-sole-make-brits-eat-it

Joker. (2019). *Feature film*. Directed by T. Phillips. Burbank, CA: Warner Bros Pictures.

'The Laughing Fish'. (1992). Batman: The Animated Series. In *Batman: The Complete Animated Series* [Blu-ray, B07FSRBWXK]. Burbank, CA: Warner Bros Pictures.

Lewis, J., Williams, A., and Franklin, B. (2008). A compromised fourth estate? *Journalism Studies*, 9(1), 1–20, DOI: 10.1080/14616700701767974

Morgan, K. and Cole, M. (2011). The discursive representation of non-human animals in a culture of denial. In R. Carter and N. Charles (eds), *Humans and Other Animals: Critical Perspectives*. London: Palgrave Macmillan.

Noske, B. (1999). *Humans and Other Animals: Beyond the Boundaries of Anthropology*. London: Routledge.

Stewart, K. and Cole, M. (2009). The conceptual separation of food and animals in childhood. *Food, Culture and Society*, 12(4), 457–476.

10 Dystopian or utopian fiction?

The sociological imagination and the
representation of pandemic futures
in *The Animals in That Country*

Josephine Browne

Introduction

This chapter discusses the Australian novel, *The Animals in That Country*, by
Laura Jean McKay (2020), released during the first lockdown of the COVID-19
pandemic in Melbourne, in March 2020. McKay's novel – her debut – is read as
speculative dystopian fiction, prescient in its preoccupation with issues of pandemics and human relations with other animals. The novel has two main characters: Jean, a middle-aged alcoholic woman who works as a guide in a northern
Australian wildlife park, and Sue, a seven-year-old dingo who has known Jean
since she and her brothers were found 'curled like beans under a bit of tin'
(McKay, 2020, p. 5).

Drawing on the cultural sociological examination of the usefulness of novels in furthering sociological understandings and aims and the relative neglect
of dystopias (Seeger and Davison-Vecchione, 2019), along with calls to make
greater use of them (Thaler, 2021), I consider McKay's (2020) work in the context of the sociological imagination (Mills, 1959). In particular, this novel is a
useful tool for linking biography and history, including the biographies of species other than humans. I argue that McKay's text can be read as valuable in
sociological terms, distinct from literary (Milner, 2016) by making visible what
Cudworth (2016, p. 249) terms 'anthroparchy' – that is, 'sets of relations of
power and domination, which are consequential of normative practice', with
attention to 'other complex forms of domination (such as patriarchy, capitalism and colonialism)'.

In considering this conceptualization of anthroparchy from the perspective
of both main characters – a human and a dingo – the novel troubles the expectations of the genre *dystopia*. Ultimately, I argue, a sociological reading of this
novel reflects back on the reader's own positionality. This reflection reminds
the sociologist of Mills' (1959) insistence on a thorough and ongoing understanding of their own position within history, acknowledging socialization and
inevitable limitations, 'always critically interrogating how our innate biases and
structural location may impact' (Fernandes, 2021, p. 10), while critiquing a
'focus solely on *man*' that 'maintain[s] the centrality ascribed to human reason
in the social sciences' (Kemple and Mawani, 2009, p. 242). By incorporating a

DOI: 10.4324/9781003257912-14

non-human main character in Sue, *The Animals in That Country* moves beyond human-centric paradigms to consider the subjectivities of other species, providing complex readings for sociologists – particularly scholars and students of multispecies sociology.

Situating the text sociologically

> No social study that does not come back to the problems of biography, of history and of their intersections within a society has completed its intellectual journey.
>
> (Mills, 1959, p. 6)

When C. Wright Mills published *The Sociological Imagination* in 1959, it was a summation of much of his former sociological work (Dandaneau, 2008), as well as an urgent call to centralize 'the promise' of sociology as a discipline for society's good (Mills, 1959). The imaginative link between biography and history he was invoking has parallels in the imaginative worlds of fiction: indeed, Mills himself argued that writers of fiction, 'whose serious work embodies the most widespread definitions of human reality ... frequently possess this imagination' (Mills, 1959, p. 14).

Along with other cultural products, literature has frequently been used by sociologists to reflect on social worlds, teach students sociological ideas and advance public sociology (Watson, 2021), which recognizes the foundations of the discipline in its close links to literary criticism (Váňa, 2020). Thamala Olave (2021, p. 418), however, argues that 'the subjective significance and existential impact of ... reading has yet to be explored within sociology', which is particularly important for multispecies sociologists in relation to McKay's (2020) text – a point to which I will return later. There is nevertheless an established understanding of 'the mutual relations between sociological imagination and literary sensitivity' (Jayaram, 2019, p. 133), demonstrating why the term 'literary sociologists' (Váňa, 2020, p. 185) has been suggested as appropriate for particular writers.

Mills' (1959) work has often been drawn upon to justify the inclusion of utopian fiction within sociology. This argument is supported by Mills' central concern for sociologists to embrace their moral purpose within the discipline (Flanagan, 2016), applying sociological knowledge in order to imagine increasingly equitable future societies, while being 'especially mindful of the dialectics of agency and structure' (Dandaneau, 2008, p. 385). Utopian fiction, historically understood as narratives associated with optimistic idealism, 'fetishiz[ing] the impossible and unachievable' (Nersession, 2017, p. 92) seems, ostensibly, to align with the sociological imagination.

However, this elides the frustrated and political nature of Mills' work, particularly in his claim that the 'promise' of the sociological imagination was under threat of being unrealized, stagnating within 'the personal uneasiness of individuals ... and the indifference of publics' (Mills, 1959, p. 5). Mills' use of the words 'imagination' and 'promise' was in the service of a specifically critical sociology, where the 'indifference of publics' can be 'transformed into involvement with public issues' (Mills, 1959, p. 5), distinct from the benign association of these words outside the discipline. As Kemple and Mawani (2009, p. 228) argue, Mills aimed 'to expand a politically aware, self-reflective and publicly accessible intellectual culture', where links between biography and history were more readily made: 'The sociological imagination enables its possessor to understand the larger historical scene in terms of its meaning for the inner life and the external career of a variety of individuals'. Mills' theory, therefore, has also been described as a concern that a postmodern dystopia may overtake the world without the intervention of sociologists: Mills is appalled by 'the reproduction of mass society by means of mass society' (Dandaneau, 2008, p. 385); by changes in modernity's efficacy that result in frequently invisible and/or justified practices of domination, and by the increasing power of media to shape post-war societies (Mills, 1959). Due to these critiques, Mills specifically cautions against ignoring the present realities in favour of focusing on idealized futures, urging sociologists to ensure they gain a thorough understanding of 'this particular society as a whole' (Mills 1959, p. 6).

The realization of sociology's 'Promise' (Mills 1959) is under threat because of insufficient consideration of two vital domains: first, the sociologist's capacity for thorough-going reflexivity, honestly appraising their positionality; and second, the sociologist's capacity to comprehensively link biography and history with absolute clarity, and with courage. Recentralizing Mills' critical and political aims within the promise of the sociological imagination supports countering sociology's tendency to conflate his ideas with utopic visions. A recent argument made by Seeger and Davison-Vecchione (2019), drawing on Mills, additionally foregrounds the research basis of dystopic fiction as an argument for its expanded inclusion in examinations of the future by sociologists. This is reinforced by Thaler's (2021) argument that the aims of multispecies justice might be furthered with a greater focus on the dystopian, which he characterizes as, '*If-This-Goes-On*', drawing on African-American writer Octavia Butler's formulation.

Acknowledging a history of sociology's fascination with utopias – for instance, in the work of Levitas, Moylan and Bauman – Seeger and Davison-Vecchione (2019, p. 51) argue that dystopias may be more useful as sources for sociological analysis, seeking to 'revive the notion of speculative literature as a form of sociology in its own right'. Dystopias are valuable, they argue, because of their fact-based critique regarding current problems in human societies and their tendency to be narrated using the point of view of a protagonist from the present world experiencing a dystopic journey as a *participant* in a particular society (Seeger and Davison-Vecchione, 2019). This immersive, subjective experience is differentiated from utopias, in which a character usually experiences a world as an outsider, a visiting observer (Seeger and Davison-Vecchione, 2019). Moreover, the nature of the interplay between the individual's biography and the wider historical structures – the focus of the sociological imagination, according to Mills (1959) – is the central concern of dystopic novels. This furthers the sociologist's aim for a continually expanding reflexive attention to our own positionality in relation to historical realities. McKay's (2020) dystopia provides this critical lens by focusing on complex relations of both affective and instrumental human relations with a wide range of other species. These irruptions within the text support an expanded vision for sociologists, enriching the disciplinary concern by rendering invisible and normative social institutions and practices visible.

The Animals in That Country (McKay, 2020) is additionally situated in a dialogue (or acknowledgement) with a poetry collection by Margaret Atwood (1968) – in particular, Atwood's titular poem, 'The Animals in That Country'. While it is beyond the scope of this chapter to do justice to the complex colonial histories invoked by the dialogue, McKay's (2020) attention to postcolonial critique requires foregrounding as a reminder of 'the tendency to universalize European and American history and thought' (Kemple and Mawani, 2009, p. 229), which 'hides or ignores the embodied stance of the spectator' (Kemple and Mawani, 2009, p. 231).

Atwood's (1968) poem, 'The Animals in That Country' is retrospective, political and mournful: it contrasts the valuing and placement of animal-others in Indigenous Canadian epistemologies with their devalued instrumental relations within colonialist practices and politics. Atwood's (1968) poem, then, compares the present antithetically with a past when 'animals had faces', in a conceptualization of shared personhood and the consequent equally sacred value of all life in Indigenous Canadian cultures prior to colonial invasion (Martin and Garrett, 2010). Now animals are worthless, 'a flash of eyes in the headlights' (Atwood, 1968), reduced to their utility value by colonial violence – both ontological and material. Atwood's poem articulates sorrow for a lost utopia in Canada's past, 'the tribal web of kinship rights and responsibilities that link the people, the land, and the cosmos together in … system[s] of mutually affecting relationships' (Justice, cited in McKegney, 2009, p. 207). McKay's world of 'that country' (Atwood, 1968), however, shifts away from the past to the present and the future, and from Canada to Australia, beginning on the land of the 'Kungarakan people' (McKay, 2020, p. 6), south of Darwin in the Northern Territory. These themes, linking the poem and the novel, are aligned with the work of post-colonial scholars interrogating a 'decolonization of thought … in a substantive and structural and physical way' (Todd, 2016, p. 17), and for ongoing sociological attention to 'how intellectual claims are implicated in the dynamics of empire-building' (Kemple and Mawani, 2009, p. 229). Colonization is a current, rather than a past, reality (Todd, 2016), and Kemple and Mawani (2009, p. 230) argue that 'the sociological imagination needs to be expanded to account for non-human forces … displacing its anthropocentric orientation'. Such challenges are a powerful way to frame possibilities that adequately respond to, and contest, the dominant crises of human–animal relations and climate and are a reminder of the significance of the decolonizing process that has been conceptualized, but remains unrealized, within the discipline (Kemple and Mawani, 2009; Todd, 2016).

The impact of this novel may have been less were it not for the COVID-19 pandemic, understood to be zoonotic (Mallapaty, 2021), emerging through human encroachments that cause 'unanticipated encounters between animal and human habitats, [and] migration shifts of animals' (Sansonetti 2020). To further situate this novel sociologically within 'the process of meaning-making … in a given socio-historical setting' (Váňa, 2020, p. 180), where 'production and reception shape the landscape of meaning' (Váňa, 2020, p. 187), it is useful to consider the context within which this dystopian pandemic text was published.

In Australia, the outbreak of the pandemic closely followed a period of significant climate distress in the Black Summer of bushfires 2019–2020, where for the first time fires generated their own weather systems, dominated the whole east coast of the country and burned for longer than ever before in our history (Canadell et al., 2021; Mullins, 2021). Three billion animals are estimated to have died, and the grief for more-than-human victims and survivors of a range of species – although frequently reinscribing hierarchies (O'Sullivan, 2020) – was significant, not just in Australia, but worldwide (Celermajer, 2021). These fires led to a change in the conceptualization of the usual Australian 'bushfire season' and are widely considered to have contributed to altering attitudes towards the climate crisis and the need for urgent action from the Australian government (Mullins, 2021). At the time of writing, as we approach another bushfire season, Black Summer remains the reference for Australia's (un)preparedness (Turton, 2020), along with broader policy failure at COP26 (Worrall, 2021).

The situating of McKay's novel within the sociocultural milieu of its reception is furthered by its receipt of the highest monetary award for literature within Australia, the Victorian Premier's Literary Award (2021), and a short-listing for the Stella Prize for Australian women and non-binary writers (2021), as well as the ABIA Small Publishers Adult Book of the Year (2021). In September 2021, McKay additionally received the (American) Arthur C. Clarke Award for Speculative Fiction. While the complex politics of awards are beyond the scope of this chapter (Jacob and Viswanatha, 2018; Kidd and Thomas, 2019), this recognition for *The Animals in That Country* (McKay, 2020) speaks to the sociocultural resonance of this work within the publishing context of both a zoonotic global pandemic and broader environmental anxieties regarding climate crisis. This also speaks to McKay's (2020) successful construction and communication of an effective speculative imaginary, reflected in laudatory reviews (Armstrong, 2020; Barnes, 2021; Brooker, 2020; Browne, 2021; Jordan, 2020); indeed, a number of reviewers admit being disconcerted by the non-didactic power of the text to call human–animal relations into question, such that 'in the days after finishing it, the world felt different to me, its animals not speaking, but not silent either' (Brooker, 2020). McKay herself has explained that she was a meat-eater when she began research for this novel, 'out of social habit and fear of stigma' (Braithwaite, 2020), but that the process of research and writing challenged her perspectives and changed her practices (Braithwaite, 2020).

Through Jean's gradual induction into exposing an anthroparchal world, on a road trip to find her animal-loving granddaughter, McKay (2020) effectively critiques norms that conflate affective relations with other animals with both childhood and (female) gender, links that Cole and Stewart (2016, p. 99) describe as 'a symbiotic process of the manufacture of sentiment and its infantalization-feminization'. The social imperative of 'maturing' into norma-tive relations with other animals involves a complexity of rendering human violence – such as mass incarceration and slaughter (Cudworth, 2015, 2016; Wadiwel, 2015) – invisible and acceptable, while affirming a 'love' of (certain) animals, 'ownership' of pets and the 'normality' of consuming the infant body parts of other species (Cole and Stewart, 2016).

Jean's world is dystopian precisely because her experience expands her understanding of her own enmeshment within Western socially normative instrumental relations with other species: she is exposed to links between her own biography and history, and these are unsettling because she has not con-sidered human–animal relations critically before. For humans, McKay's pan-demic of 'zooflu reveals the *dis*harmony, fear and pain that we are surrounded by but have been too self-centred to acknowledge' (Laird, 2021, p. 34). Thus, through a range of engagements with other species, Jean's experiences are sociologically useful in considering a range of anthroparchal relations of dom-ination (Cudworth, 2016), including their ontological construction.

In contrast, Sue's view as a dingo indicates that other animals are already communicating, including with humans: Sue is sanguine about the new human ability to dialogue with other species. The inclusion of Sue's subjec-tivity, as rich and nuanced as Jean's, along with other species' subjectivities, prompts a reconsideration of the 'eternally fresh and controversial' (Lee, 1976, p. 925) question of, 'sociology for whom?' (Burawoy, 2005), creating contention regarding whether – or for whom – McKay's text can be classified as dystopian.

Overview: *The Animals in That Country*

McKay's novel begins in a wildlife park in northern Australia. A zoonotic pan-demic – zooflu – is spreading from the large south-eastern city of Melbourne, and one of its side-effects for those infected is the ability to understand the communications of other animals. This is not through speech, but an emergent ability 'to communicate (encode) and translate (decode) previously unrecogni-sable non-verbal communications via major senses such as sight, smell, taste, touch, and sound' (McKay, 2020, p. 35). When Jean's infected son, Lee, takes her granddaughter south, following the whales (McKay, 2020, p. 108), without the permission of the girl's mother, Jean and Sue set out to find them, on a road trip across 'broken roads and orange dust of a country so big I'll never see the end of it, never find such a small girl' (McKay, 2020, p. 167).

Once infected with zooflu herself, Jean is able to understand Sue's communications, though falteringly, like the shadow of a language she once knew: 'You have to look at all their whole body all at the same time', her granddaughter Kimberly tells her (McKay, 2020, p. 91). Jean works to piece together the rich mixture of signals: 'The smell from Sue's hairy armpits ... the quiet and constant song for blood across her gums' (McKay, 2020, p. 117), 'She's speaking in odours, echoes, noises with random meanings popping out of them' (McKay, 2020, p. 81). The human interpretation of the animals' communications are rendered in the text as short, often disjointed, poetic lines, as in Jean and Sue's first encounter after Jean is infected, when Jean starts to read Sue's 'twitchy paw, the rumbles in her throat, her smooth pelt, her smart-as-a-whip ears' (McKay, 2020, p. 82): 'The body of my/mother, tang/of blood./My mother/tongue. The Queen/(Yesterday)' (McKay, 2020, p. 82). Jean recognizes Sue's superior ability to find Kimberley: 'In Sue's drool and eyelids and paws, Kim is an animal' (McKay, 2020, p. 114), her 'body sings a picture ... bruises on the girl's scrawny legs ... freckles not yet appeared on her skin, like the wee that will need to happen in an hour, like the washed sheets she slept on, and the sting of adventurous fear when she took Lee's hand' (McKay, 2020, p. 112).

The use of the road trip trope provides McKay with the opportunity to explore human–animal relations through encounters with a wide range of other species. These include animals such as pigs, 'dairy' cows, 'wildlife', birds, 'pets', 'pests', insects – and, at the novel's climax, whales, who are calling humans to 'come home' to the ocean (McKay, 2020, p. 207). The narrative draws attention to the negative impact of human animals, even on animals largely understood as 'free' – like the kangaroos killed or injured on the road by cars (McKay, 2020, pp. 152–54) and the celebrated whales, who 'splinter like someone has thrown a rock at a song' (McKay, 2020, p. 227) when the police enter the water in motorized boats.

The reader is exposed to an increasing range of other animal communications through the subjectivity of the human character, Jean, as a member of a society and country gripped by the pandemic (Seeger and Davison-Vecchione, 2019). Jean is a tour guide because she failed a test to be a ranger. She hides her drinking habit so she can have access to her granddaughter, whose mother runs the wildlife park. It is significant that Jean is socially situated within this text as an 'animal lover', living in a staff cabin in the park with a range of 'rescued' animals unfit for exhibiting, and dreaming of building a sanctuary with her granddaughter; '*Come to Kim and Granny's Animal Place. No animal turned away*' (McKay, 2020, p. 161). Jean and Kimberly agree that they want the flu (McKay, 2020, p. 37) – they have always loved animals and want to talk with them. In the absence of shared speech, Jean has created voices for other animals, including during her guided tours of the wildlife park with visitors: the dingoes bicker, she claims, 'Give me that chow, you mongrel' (McKay, 2020, p. 6), and a captive barking owl calls, 'Have you seen some mice around here?' (McKay, 2020, p. 18).

Through the other main character in this novel, Sue, McKay (2020) also situates the reader to consider how other animals might appraise humans. As a dingo, Sue is sanguine about Jean's emerging ability to understand both her and the other animals around them: there is no reaction or comment on the shift in communication. This absence speaks to the normative of the other-animal world in which Jean finds herself, and she is disgruntled to discover the other animals are not living the human-centric lives she had imagined – a normative hierarchical imaginary she was actively passing on to Kimberly. Sue and all the other animals have always been communicating: the communication difficulties in this novel are situated within the human species, whose selective ignoring of the communicating more-than-human world has been normalised, along with imposed ontological categories, motives and presumed speech for (certain) other animals. As Jean's infected son, Lee, tells her, 'They're always saying stuff. We just, you know, didn't get it' (McKay, 2020, p. 65).

Jean is confronted with received human arrangements of normative animal hierarchies and exposure to her own animality in her journey with Sue, who claims from the outset that she and Jean are 'kin' (McKay, 2020, p. 150). Late in the novel, their relations inverted, Jean relies on Sue's ability to keep them alive, food and petrol having run out, as they make their way home on foot:

> Sue can tell if water is cleanish, and how long things have been dead. She can hear the heartbeats of the mice and rabbits hiding underground … she knows where the eggs are, how to root around for things buried in the cold dirt, how to sleep curled around each other.
>
> (McKay, 2020, p. 260)

Sue has demoted Jean, now calling her 'Good/cat' and 'A Bad/Dog' (McKay, 2020, p. 261), when she objects to eating 'another putrid mouse' (McKay, 2020, p. 261). Jean has incrementally realized her own status as an animal, and the misguided nature of her assumptions about other species, agreeing now with Sue, 'Its pack is/my/pack pack' (McKay 2020, p.204).

Presciently paralleling the COVID-19 pandemic taking place in the world in which the novel was published, McKay (2020) presents a wide range of human responses to the virus, from the government and social media to conspiracy theorists, church groups, the army and individuals. The virus intensifies its effects as it worsens for humans. Initially, their senses are inundated with communications from larger animals; at its worst, zooflu enables people to hear even the minute insects within the bark of trees and in the earth. This cacophony of previously ignored communications is represented as driving many humans to madness, and they respond by using face masks and ear plugs, and even drilling holes in their heads, in an effort to stop deciphering the other animal worlds by which they are surrounded.

The denouement of McKay's (2020) novel is undoubtedly bleak: Jean is threatened by roving army personnel to take the zooflu antidote. Sue 'looks up at me, swallows', asking, 'Is this its/friend' (McKay, 2020, p. 272). Jean, failing to vomit the pill back up, struggles against the impact of the loss of her other-animal connections, above all with Sue: 'wait[ing] for the meanings to dribble off her tongue … The universe gone' (McKay 2020, pp. 276–77).

Dystopias, utopias and multispecies sociology

The Animals in That Country, when read with attention to Mills' (1959) analysis linking biography and history within the sociological imagination, as well as a critique that queries 'dystopia for whom?' is a text rich in examples of Cudworth's (2016) conceptualization of anthroparchy. In the current crises of climate change and associated pandemics and mass extinctions, historic notions of utopias as a place where human need is finally satiated might better be retrospectively (re)read as the height of dystopia (Nersession, 2017). After all, 'success' for capitalist ideology is constructed on the necessity of manufacturing endless human needs and wants. Applying an intersectional feminist analysis to the construct of dystopia, along with a critical sociological suspicion of dialectics, exposes implied alternatives present within McKay's (2020) text.

Thus far, I have suggested that *The Animals in That Country* (McKay, 2020) can be read as a pandemic dystopia, 'at times … a post-apocalyptic horror story' (Armstrong, 2020, p. 315). Jean's subjectivity (female, mother, grand-mother, working-class, animal-loving, alcoholic, carnivorous, etc.) guides the reader's immersion into an alternate future, from a foundation of realism in contemporary Australian society (Seeger and Davison-Vecchione, 2019). McKay's decision to write the novel from Jean's first-person perspective saturates the reader with the exploration of an emergent world, read sociologically as an exposure to a range of anthroparchal relations, in which the varied situation of a range of other animals within human systems of categorization and domination – such as wildlife, tourism, industrialized agriculture, religion, government and policing – are articulated.

Situating Sue, a dingo, as the secondary main character creates a sophisticated alternative species viewpoint. First, in an adult novel with serious philosophical and ethical themes, having a main character who is another species is itself a departure that reflects increased attention to the validity of sociological critiques being articulated within the academy regarding human and more-than-human relations (Cudworth, 2016; Stewart and Cole, 2021). In addition, while reader access to Sue's viewpoint is limited by Jean's first-person narrative, and the necessity of her translations (Armstrong, 2020), the ability of Sue (and other animals) to be 'heard' by the human species remains significant. For Sue and other animals, it is humans who have arrived late to the already busy, communicating more-than-human world, and humans who have been fascinated with the possibilities of dialogue are affronted by the animal-centric continuity they discover. This, along with the representation of other species as agentic, raises implicit challenges for diminished and human-centric assumptions by exploring a range of both localized and globalized practices inflicted on other species, including affective and instrumental relations. This possibility of becoming visible and reconceptualized as agentic with both individual and species needs, desires and priorities might suggest a vision closer to utopia, from Sue's (and other animals') point of view. Her erstwhile experience has been that of a caged animal, a tourist exhibit in a 'wildlife' park. Now, her 'white chest talk[s] two ways. One for the open road, the time of the whole world, the wild dogs out there. The other way for inside the cage, the safety of locked doors and a hand on her back' (McKay, 2020, p. 81). McKay's critique of Sue's material realities as a caged native animal is only one of many species' experiences represented in this novel. Sue's point of view richly describes her individual loyalties, conflicts, preoccupations, fears and desires throughout the novel, as well as her perceptive insights into the humans around her. By the narrative's end, she is wary of the 'antidote', warning Jean, 'Watch/it (Cat Dog)' (McKay, 2020, p. 271).

In articulating a range of anthroparchal relations that 'assume specific spa-tialized and historical formations' (Cudworth, 2016, p. 249), McKay's narra-tive enables a critique of the unstable ontological categories humans impose on other animals, such as wild, farmed, pets, feral and native. Sue is represented as 'wildlife', a tourist commodity, but she is simultaneously situated in conflicted colonialist discourses about dingos in Australia, particularly their hybridiza-tion, 'where race, gender and species intersect' (Probyn-Rapsey, 2015, p. 55). McKay explores these conflicted positions for Sue – 'all kelpie-cross. All dingo' (McKay, 2020, p. 163) – such as when Jean offers people 'five bucks a question' (McKay, 2020, p. 178) in an attempt to make money from Sue and her situation within this nexus of Australian cultural fascination. One person asks, 'Are you wild or tame?' (McKay, 2020, p. 178), while another tells Sue she's 'not even native, you know that? You're not even from here' (McKay, 2020, p. 179). Sue complains to Jean that subjecting her to this indignity is 'a/rotten bits/plan' (McKay, 2020, p. 177). As Sue becomes increasingly objectified by the crowd, reduced to a representative of her species, 'panic waves pour off the roof of the camper' (McKay, 2020, p. 178) on which she sits, 'feel of their hands on my fur, leather around my face, tight on my neck' (McKay, 2020, p. 178). Later, Sue's complex enmeshments in conflicting discourses for dingoes are further explored in her pursuit of a mate (McKay, 2020, pp. 242–57). Sue's experience as a dingo in Australia offers a critique of notions of an idyllic, utopian, former purity to which 'we' might return, a conceptualization that dominates metanarratives of the Anthropocene (Nersession, 2017). The pursuit of such nostalgic returns avoids the very complex historical situation of a range of animal species, justi-fying resulting efforts to eradicate 'impurity', and has intersectional implica-tions within colonialism, racism and speciesism (Mayes, 2020).

McKay (2020) further exposes anthroparchal relations in Jean's encounters with animals subjected to industrialized farming practices – cows to produce 'dairy' and pigs for 'meat'. In the former instance, Jean is recognized by the cows – Lee's father, her former partner, still lives on the farm. Pushing against a gate among 'smells like milk, fermenting grass, grief and heat' (McKay, 2020, p. 182), the cows address her: 'It came and/made babies from/babies. Where/are/they' (McKay, 2020, p. 182). When one bursts out, 'Fuck my/ clumpy teats where/are they' (McKay, 2020, p. 183), Jean rejects any acknowledgement or responsibility for her own participation in these instrumentalized relations: 'These cows want to tell me stories I don't want to hear. Got my own lost babies to find, thank you very much' (McKay, 2020, p. 183). The cows entreat Jean, on the basis of their shared positions as mothers, whose 'face ... holds children' (McKay, 2020, p. 192), raising the intersectional connections between specific practices of dominating females of other species and human women (Cudworth, 2016; Gunderson, 2014), justified by patriarchal and colonial logics: 'difference, thus represented as lack ... becomes the basis of hierarchy and exclusion' (Plumwood, 2002, p. 13). After the visit exposes Jean's own mistreatment by the farmer, and he attempts to shoot Sue, Jean allows the cows' communication to impact her: 'Their bodies make pictures that get inside my bones until I'm half cow with skin shivers and nose chatters, pregnant again and again' (McKay, 2020, p. 193). Jean admits to the mother cows, 'I do know' (McKay, 2020, p. 194), as she and Sue leave, united now in interspecies female solidarity in relation to the farmer.

Eschewing didacticism, McKay's (2020) treatment of scenes such as that on the 'dairy' farm is able to demonstrate the complex engagements of human economics, gender and colonialism with instrumental intersectional histories of female cow incarceration, forced breeding, calf removal and 'slaughter', and the premature 'slaughter' of adult cows, within the industrial complex of the 'dairy' industry in modern Australia, as well as the complexity of the position of farmers, men and women, within these constructs (Taylor and Fraser, 2019). Such a micro analysis as provided by the text has transparent links to macro systems of the increasing use of dairy for non-Western nations, thus simultaneously linking the individual experience of grief and incarceration for this cow, who becomes impatient with Jean's obfuscation, with broader systems of relations of domination for both dairy cows specifically, and more implicitly, cows as a species in relation to humans (including relations of 'slaughter' and 'meat').

A final example of the many possibilities for anthroparchal exploration within the text considers human impacts on a kangaroo – an animal largely conceptualized as 'free', as well as valorized as 'wildlife' (conflicting with a thriving leather and meat 'market', see Hauser, Pople and Possingham, 2006). Jean hits a kangaroo on the road through the desert and as she gets out to apologize – an action that implicitly critiques normative responses to 'roadkill', and hearkens back to Atwood's poem (1968), which likewise does so – Jean tends to the dying kangaroo lying on the road, her 'soft brown eye like glass, blood a tear from its [sic] mouth, breathing with effort' (McKay, 2020, p. 153), and hears a 'whisper from under its skin' (McKay, 2020, p. 153). Jean finds a joey and, conflicted by travelling with Sue, who would perceive the joey as food, as well as her need to find her granddaughter, reluctantly wrings the joey's neck: 'I lay the little roo, very slow, next to the big one' (McKay, 2020, p. 154).

This scene connects to others in the book where wallabies and kangaroos are situated, both in the wildlife park as commodities, and with a wallaby Jean lives with, Wallamina, another 'rescue' from a mother killed on the road. Wallamina spends her days fixed in a corner of Jean's yard, and it is not until after the advent of zooflu that Jean, who has previously fondly disparaged the wallaby as crazy, realizes Wallamina thinks her mother is there (McKay, 2020, p. 66).

The critique of the concept of 'wildlife' as a commodified spectacle is further extended when Jean discovers 'some sort of animal sanctuary' (McKay, 2020, p. 161) – a money-making venture for tourists – next to an outback pub. The animals are locked in cages and dying, and the publican encourages Jean to use alcohol to avoid hearing their communications: 'If you drink enough, you can't hear them calling' (McKay, 2020, p. 163). An anthroparchal framing of these scenes situates the treatment of these animals in relation to humans in complex and conflicting ways. Among these are the ontological categorization of 'native', incursion of human structures and transport and the consequent destruction or displacement of other animals and their homes, as well as the generic, reductive normalization of other animal deaths in car accidents, labelled 'roadkill'. McKay (2020) additionally demonstrates human-imposed hierarchies of relative value for other animals in the 'rescuing' of (only) those labelled 'native' from road accidents and the instrumental valuing of 'wildlife' for tourism and, by extension, for 'protection'. In the textual focus on two individual joeys left motherless due to the incursion of cars, the reader is situated to effect the links between each joey biography – revelatory within the constructs of the speculative – with a range of histories invoking human impact within the worlds of these species. This, along with Sue's dingo perspective, further displaces the human as the implicit site of 'biography' within the sociological imagination (Mills, 1959), so advancing the construction of alternate histories (for instance, human construction, transport, entertainment, tourism, etc.) from the perspective of other animals.

Jean makes both a literal and a figurative journey in this novel. Her figurative journey is into the social realities of other species as they are situated within anthroparchal relations with humans. Jean is by turns disbelieving, horrified or denying the experiences of other animals as she increasingly realizes their situation (and her own) within complex relations with human systems, and her question to a nurse who disparagingly says she is probably still eating animals – 'What the hell else am I supposed to eat?' (McKay, 2020, p. 176) – expresses her continued situatedness in normative practices of instrumental relations and consumption. The reader moves with Jean through a familiar world she is rediscovering as she attends to other-animal communications. The shift for Jean is from the world she thought she knew, and her familiar social identity as a carnivorous 'animal lover', with confrontations regarding other animal experiences within this received system of normative anthroparchy.

Jean has no overarching epiphany that alters her own behaviour; her encounters are an aside to a focus on her fears for her granddaughter, notwithstanding her deepening and significant relationship with Sue. When the nurse concludes that her case of zooflu is mild, and suggests she try 'shutting up enough to let your ego explode. Have it rebuilt by a bird, an ant, a *dingo*' (McKay, 2020, p. 176), Jean rejects this as ridiculous, reaffirming normative practices, despite all she is witnessing. This confirms other suggestions in McKay's (2020) text that it is possible for people to 'unknow' cruel and unjust practices, a possibility contested in research into veganism (McDonald, 2000).

Jean and Sue share a complex relationship, which becomes more significant once Jean develops zooflu and can understand Sue's communications. This commences with Jean upbraiding herself for earlier infantalized 'languaging' she has employed in relation to other species: '*[p]at pat walkies* had so much damned power yesterday' (McKay, 2020, p. 83). Jean's comforting claims to 'love' other animals are continuously tested against the experiences of animal others, positioning both Jean and the reader reflexively in relation to normative human practices that ideologically sanction the imposition of hierarchies of violence on other species and their habitats. The efficacy of this reader discomfort has been remarked upon by reviewers (Barnes, 2021; Brooker, 2020), as noted earlier. Jean's experiences, through this revelatory journey of 'instrumental-rational relations with other animals' (Cole and Stewart, 2016, p. 31), expose incongruities, inconsistencies and disjuncts within normalized Western, capitalist claims regarding human and other animal relations, the lack of 'a unifying logic' described by Cole and Stewart, and the subsequent 'ongoing process of legitimation of this precarious ordering as "proper"' (Cole and Stewart, 2016, p. 56). The response of some humans is to release animals; for others, it is to block the knowledge out with masks, ear plugs, alcohol or even drilling a hole in their heads. As a Christian tells Jean, 'I haven't looked. Or listened' (McKay, 2020, p. 221).

In considering the effects of zooflu on infected humans, *The Animals in That Country* (McKay, 2020) can also be read as articulating a reality that includes the consciousness of current anthroparchal relations for those who already hold a critical position regarding dominating human-centric narratives and practices. McKay's (2020) work explores an existent critical multispecies worldview (acknowledging that this is not homogenous) through the attention humans give to other species once they are infected by zooflu. In doing so, the novel speculates on the ramifications of a whole society of humans becoming newly conscious of our relations with other animals in ways akin to how those with multispecies and post-humanist concerns do within their lives and work, including activism. Situating the reading from a multispecies sociological viewpoint, McKay's (2020) text explicates present 'realities', and may therefore be read with a sense of the relief of recognition (although the text is in no way didactic), one that participates in conversations already taking place within the public realm (Lockwood, 2012), as the recognition of its awards suggest. The limit to humans living in kinder ways with other species, this dystopia suggests, is not elided by arguments touting human exceptionalism based on the reductive simplicity of a shared language, but rather humans' own 'colonising conceptual sieve that eliminates certain communicative possibilities and dialogical encounters' (Plumwood, 2002, p. 22). McKay (2020) implicitly interrogates the human childhood wish that other animals could talk, querying whether we really want to hear what their perspective on the human species might be. The novel additionally affirms there is no 'lack' in humans sharing their own form of language that creates barriers between species. The problem, which aligns with the scholarship of multispecies sociology and critical animal studies more generally, is in a human refusal to attend to the already present communications of others, 'a certain kind of deafness' (Plumwood, 2002, p. 22).

The Animals in That Country (McKay, 2020) invites the reader to reflexively consider their own positionality in encounters with multispecies positions by making the marginal visible via the subjectivity of Sue and a range of animal-others. The text is ambiguous, in keeping with the genre of literary fiction, so encouraging reflexive engagement with speculative concepts including the anthroparchal present, dystopia and suggestive utopic possibilities inherent in humans attending to more-than-human sensibilities, particularly Sue's. The text thus provides the sociologist with an opportunity to reflexively engage, 'the self-awareness generated by sociology reflecting on itself' (Flanagan, 2016, p. 18). This reflexivity, expanded by critical postcolonial and intersectional approaches (Fernandes, 2021; Kemple and Mawani, 2009; Todd, 2016), enables sociological readers to examine their own biographies in relation to this historic moment, humbly acknowledging 'there is no way ... [to] avoid assuming choices of value and implying them in his [sic] work' (Mills, cited in Flanagan, 2016). McKay's (2020) *The Animals in That Country* provides a reflexive space supporting serious consideration of the arguments for an expansion of critical sociology to include other animals (Taylor and Sutton, 2018); as Cudworth (2016, p. 247) avows, 'it is via our attentive observation and our compassion, even for creatures who might appear alien to us, that we might enter into "dialogue"'.

Conclusion

> I'm going ... to tell her that she can live two ways – the animal way and the people way.
>
> (McKay, 2020, p. 176)

This chapter has discussed the usefulness of McKay's (2020) pandemic novel *The Animals in That Country*, released in Australia during the COVID-19 pandemic, for sociologists generally and for multispecies sociology in particular, in critiquing normative anthroparchal relations. It has situated the text within established arguments for literary contributions to teaching and theorizing within sociology, particularly engaging with the notion of the sociological imagination, which Mills (1959) has argued novelists frequently possess and sociologists must continuously cultivate.

The sociological analysis of McKay's novel has been framed by Seeger and Davison-Vacchione's argument (2019) regarding the sociological usefulness of dystopian narratives, an expansion of the '*If-This-Goes-On*' (Thaler, 2021) critique, thereby contributing further research into the possibilities of speculative fiction. While utopian fiction has frequently included vegetarian diets in imagined futures, particularly feminist work (Gilman, 1979 [1915]; LeGuin, 1974), these novels have stopped short of a thorough-going critique of the animal–industrial complex. Indeed, in a more recent dystopia, *The Road* (McCarthy 2006), the height of horror is represented as the eating of human children, a suggestion of the extent of human cognitive dissonance regarding contemporary relationships between ourselves and other animals.

In *The Animals in That Country* (McKay, 2020), Jean is socially situated as an animal-loving carnivore, with a history of fractious relationships with people and a tendency to alcoholism. Jean's immersion into a dystopic world, by way of the side-effects of a zoonotic pandemic, is a journey into what I have argued is an anthroparchal worldview, one situated in Australia's present, and a speculative near-future, rather than in the past that Atwood invokes in her mourning poem of the same name (Atwood, 1968). Jean, often inebriated and distracted in her search for her granddaughter, is exposed to evidence of the instrumental relations of humans and other animals. Her subjectivity provides a human view of the complex networks of more-than-human suffering, including distortions of affective relations and linguistic diminishment, which can usefully be considered within networks of power established in Cudworth's (2016) notion of anthroparchy.

Sue, a dingo, provides a viewpoint in the novel from the perspective of another species, with her own biography and agency, her unique motives for joining the road trip and a character as complex as her human co-traveller. In representing Sue as providing no comment on the addition of humans into already-present conversations among species, and in the exposure of human centricity in the assumptions about other animals that humans routinely make, McKay (2020) invokes decolonial themes and hints at the utopic possibilities, for other species, of being attended to by humans. A wide-ranging critique of the systemic dominion of humans, as well as the effects of acts of power on individual animals, is made throughout the text in its inclusion of multiple species. In exposing and exploring other-animal subjectivities throughout the narrative, McKay's (2020) text unsettles norms by displacing human biography, usefully challenging the human-centric linking of biography and history (Mills, 1959) and providing further critique for sociological interest in the dynamics of agency and structure (Petray and Pendergrast, 2018).

I have argued that McKay's (2020) vision of dystopia is a world that already exists for those attuned to it, including the colonised and Indigenous peoples her title invokes (Simmons, 2019), and those concerned about 'humanity's irrational domination of the animal world' (Gunderson, 2014, p. 285). The novel can be read as explicating human failure to listen, and the necessary imposition of a silenced 'other' (Plumwood, 2002) in order to continue reproducing normative practices of human domination of all other species, and elide the instabilities of contradictions and fissures (Cole and Stewart, 2016). The silence of other animals, this reading of McKay (2020) affirms, is structural and deliberate rather than natural; it exists to enable the continuance of contemporary human–other-animal relations and to obscure the reality of humans themselves as animals. We might also link this to the continued silencing of critiques of the COVID-19 pandemic and climate crises that take seriously these warnings that anthroparchal relations cannot continue unchanged.

Jean's response after she is forced to take the zooflu antidote – 'my empty ears, my empty nose, and my eyes' (McKay, 2020, p. 277) – suggests what an absence of this attention to other animals leaves: 'The universe gone' (McKay, 2020, p. 277). For Jean, this loss – particularly of her ability to communicate with Sue – has become more significant than the tortures that exposure to multispecies communication (and suffering) have produced. By making the familiar strange through a norm-challenging reversal of marginal multispecies worldviews into the mainstream, readers are able to experience an alternate imaginary that exposes and explores both the structural and material realities of the human species' impact on other animals through multiple viewpoints, including those of other species. The multiple positionalities of the reader have prompted the approach of this chapter, which has queried for whom the text might be considered dystopic and for whom it gestures toward 'The Promise' (Mills, 1959).

The inability of humans to 'talk' with other animals is inverted by McKay (2020) into a critique exposing the necessity and limitations of human social norms. In other words, current arrangements between species, largely imposed by human dominance of the planet, are not consequent on a lack of communication on the part of the more-than-human. Rather, colonialism's legacy includes human failure to listen; this is exposed as a sociopolitical construction to enable continuing asymmetrical power relations. Other animals are always communicating, and it is a human conceit (painstakingly constructed as a political and social necessity) to ignore this.

Ultimately, the pandemic in *The Animals in That Country* (McKay, 2020) is a plot device to seriously explore ethical and philosophical questions of human and other animal relations that are central concerns for multispecies sociology. It slips between what it might mean if humans could communicate with other species, presenting a dystopic vision that rejects reductive representations of animals in the fiction of childhood (Cole and Stewart, 2016) and a realism suggesting that the anthroparchal world of its imaginary is already present if humans are willing to listen and attend. It also suggests a utopian future in which animal others are attended to, an extension of Atwood's (1968) concerns about alternative Indigenous epistemologies. McKay's (2020) novel encourages a reflexivity in the reader regarding their own biography in relation to anthroparchal histories. The narrative's efficacy in promoting reflection on human–animal subjectivities and relations has been commented on by reviewers (Barnes, 2021;

Brooker, 2020). The awards the work has garnered are an indication of its sociocultural resonance (Váňa, 2020) in its publication during a pandemic, and in the wake of Black Summer bushfires in Australia, as well as increasing public concern about the environment, and for human treatment of other animals (Celermajer, 2021).

Through considerations of biography and history raised within the sociological imagination, dystopia and utopia as species-relative and fluid, and the notion of anthroparchy, this chapter has suggested a range of explorations regarding the ways *The Animals in That Country* (McKay, 2020) engages with rich possibilities for sociological analysis, particularly for those interested in multispecies research. Further scholarship might consider the novel within the Indigenous cosmologies to which its title alludes, in referencing the work of Atwood (1968) or in advancing the examination of the subjective and existential experiences of readers (Thamala Olave, 2021). McKay's (2020) pandemic dystopia confirms the strength of sociological imaginaries that link the public and private, particularly in contributing serious attention to the place of multispecies biography. *The Animals in That Country* (McKay, 2020) justifies continuing the rich historic links between literature and sociological analysis, as well as specifically furthering critiques advancing multispecies sociology, in order that the 'Promise' be kept alive (Mills, 1959).

References

Armstrong, P. (2020). The animals in that country by Laura Jean McKay: Review of the animals in that country by Laura Jean McKay. *Animal Studies Journal*, 9(2), 311–18.

Atwood, M. (1968). *The Animals in That Country*. New York: Little, Brown.

Barnes, S. (2021). *Reading the Cues: Review of The Animals in That Country*, by Laura Jean McKay. Sydney Review of Books. https://sydneyreviewofbooks.com/review/mckay-the-animals-in-that-country

Braithwaite, E. (2020). Laura Jean McKay: On speculative fiction, veganism, our relationship with nature, and her novel. *The Animals in That Country*. www.writeordietribe.com/author-interviews/interview-with-laura-jean-mckay

Brooker, B. (2020). *The Animals in That Country* by Laura Jean McKay: Review of The Animals in That Country, by Laura Jean McKay. *Australian Book Review*, 421. www.australianbookreview.com.au/abr-online/archive/2020/may-2020-no-421/787-may-2020-no-421/6450-ben-brooker-reviews-the-animals-in-that-country-by-laura-jean-mckay

Browne, J. (2021). Speculative pandemic challenge to human exceptionalism: Review of The Animals in That Country, by Laura Jean McKay. *TEXT Journal of Writing and Writing Programs*, 25(2), 40–46.

Burawoy, M. (2005). For public sociology. *American Sociological Review*, 70(1), 4–28.

Canadell, J.G., Meyer, C.P., Cook, G.D., Dowdy, A., Briggs, P.R., Knauer, J., Pepler, A. and Haverd, V. (2021). Multi-decadal increase of forest burned area in Australia is linked to climate change. *Nature Communications*, 12, Article 6921, pp. 1–11.

Celermajer, D. (2021 *Summertime: Reflections on a Vanishing Future*. Ringwood: Penguin Random House.

Cole, M. and Stewart, K. (2016). *Our Children and Other Animals: The cultural construction of Human–Animal Relations in Childhood*. London: Routledge.

Cudworth, E. (2015). Killing animals: Sociology, species relations and institutionalized violence. *The Sociological Review*, 63, 1–18.

Cudworth, E. (2016). A sociology for other animals: analysis, advocacy, intervention. *International Journal of Sociology and Social Policy*, 36(3/4), 242–57.

Dandaneau, S.P. (2008). The sociological theory of C Wright Mills: Toward a critique of postmodernity. In H. Dahms (ed.), *No Social Science Without Critical Theory: Current Perspectives in Social Theory*, 25, 383–401.

Fernandes, S. (2021). Storytelling. SO FI ZINE, 9, 7–10. online https://sofizine.com/latest-edition/edition-9

Flanagan, K. (2016). *Sociological Noir: Irruptions and the Darkness of Modernity*. London: Routledge.

Gilman, C.P. (1979 [1915]). *Herland*. New York: Pantheon.

Gunderson, R. (2014). The first-generation Frankfurt School on the animal question: Foundations for a normative sociological animal studies. *Sociological Perspectives*, 57(3), 285–300.

Hauser, C., Pople, A. and Possingham, H. (2006). Should managed populations be monitored every year? *Ecological Applications*, 16(2), 807–19.

Jacob, S. and Viswanatha, V. (2018). Gender and Indian Literary Awards: What do the numbers say? *Economic and Political Weekly*, 53(25), 23–27.

Jayaram, N. (2019). Sociological imagination and literary sensitivity: A tribute to M.N. Srinivas. *Sociological Bulletin*, 68(2), 133–53.

Jordan, J. (2020, 7 October). *The Animals in That Country* by Laura Jean McKay Review – an extraordinary debut: Review of *The Animals in That Country* by Laura Jean McKay. *The Guardian*. www.theguardian.com/books/2020/oct/07/the-animals-in-that-country-by-laura-jean-mckay-review-an-extraordinary-debut

Kemple, T.M. and Mawani, R. (2009). The sociological imagination and its imperial shadows. *Theory, Culture and Society*, 36(7–8), 228–49.

Kidd, K.B. and Thomas, J.T. (2019). *Prizing Children's Literature: The Cultural Politics of Children's Book Awards*. London: Routledge.

Laird, T. (2021). 'Zoognosis: When animal knowledges go viral. Laura Jean McKay's *The Animals in That Country*, contagion, becoming-animal, and the politics of predation. *Animal Studies Journal*, 10(1), 30–56.

Lee, A.M.C. (1976). Presidential address: Sociology for whom? *American Sociological Review*, 41(6), 925–36.

LeGuin, U. (1974). *The Dispossessed*. New York: Harper & Row.

Lockwood, A. (2012). The affective legacy of Silent Spring. *Environmental Humanities*, 1, 123–40.

Mallapaty, S. (2021, 16 September). Did the coronavirus jump from animals to people twice? *Nature*. https://www.nature.com/articles/d41586-021-02519-1

Martin, K.J. and Garrett, J. (2010). Teaching and learning with traditional Indigenous knowledge in the tall grass plains. *The Canadian Journal of Native Studies*, 30(2), 289–314.

Mayes, C. (2020). Governmentality of fencing in Australia: Tracing the white wires from paddocks to Aboriginal protection, pest exclusion and immigration restriction. *Journal of Intercultural Studies*, 41(1), 42–59.

McCarthy, C. (2006). *The Road*. New York: Alfred A. Knopf.

McDonald, B. (2000). 'Once you know something, you can't not know it': An empirical look at becoming vegan. *Society & Animals*, 8(1), 1–23.

McKay, L.J. (2020) *The Animals in That Country*. Melbourne: Scribe.

McKegney, S. (2009). 'Beautiful hunters with strong medicine': Indigenous masculinity and kinship in Richard Van Camp's *The Lesser Blessed. The Canadian Journal of Native Studies*, 29(1/2), 203–27. www3.brandonu.ca/cjns/29.1-2/14McKegney.pdf

Mills, C.W. (1959). *The Sociological Imagination*. Oxford: Oxford University Press.

Milner, A. (2016). Resources for a journey of hope: Raymond Williams on utopia and science fiction. *Cultural Sociology*, 10(4), 415–30.

Mullins, G. (2021). *Firestorm: Battling Super-charged Natural Disasters*. Ringwood: Penguin Random House.

Nersession, A. (2017). Utopia's afterlife in the Anthropocene. In U.K. Heise, J. Christensen and M. Niemann (eds), *The Routledge Companion to the Environmental Humanities*. London: Routledge.

O'Sullivan, S. (2020, 9 January). The animals we Rescue, and the animals we don't. *ABC Religion & Ethics*. www.abc.net.au/religion/australia-fire-crisis-the-animals-we-save-and-those-we-dont/11856714

Petray, T. and Pendergrast, N. (2018). Challenging power and creating alternatives: Integrationist, antisystemic and non-hegemonic approaches in Australian social movements. *Journal of Sociology*, 54(4), 665–79.

Plumwood, V. (2002). Decolonising relationships with nature. *Pan*, 2, 7–30.

Probyn-Rapsey, F. (2015). Dingoes and dog-whistling: A cultural politics of race and species in Australia. *Animal Studies Journal*, 4(2), 55–77.

Sansonetti, P.J. (2020). COVID-19, chronicle of an expected pandemic. *EMBO Molecular Medicine*, 12(5). https://doi.org/10.15252/emmm.202012463

Seeger, S. and Davison-Vecchione, D. (2019). Dystopian literature and the sociological imagination. *Thesis Eleven*, 155(1), 45–63.

Simmons, K. (2019). Reorientations; or, an indigenous feminist reflection on the anthropocene. *JCMS*, 58(Winter), 174–179.

Stewart, K. and Cole, M. (2021). British innovations in vegan sociology. Paper presented to the Vegan Promise: Vegan Sociology as a Conduit for Human and Non-human Emancipation, International Association of Vegan Sociologists (IAVS) inaugural conference, 9–10 October. www.youtube.com/watch?v=BEr9bNHjd78

Taylor, N. and Fraser, H. (2019). The Cow Project: Analytical and representational dilemmas of dairy farmers' conceptions of cruelty and kindness. *Animal Studies Journal*, 8(2), 133–53.

Taylor, N. and Sutton, Z. (2018). For an emancipatory animal sociology. *Journal of Sociology*, 54(4), 467–83.

Thaler, M. (2021). What if: Multispecies justice as the expression of utopian desire. *Environmental Politics*, 31(2), 258–76.

Thamala Olave, M.A. (2021). Exploring the sacrality of reading as a social practice. *American Journal of Cultural Sociology*, 9, 99–14.

Todd, Z. (2016). An Indigenous feminist's take on the ontological turn. *Journal of Historical Sociology*, 29(1), 4–22.

Turton, S.M. (2020). Geographies of bushfires in Australia in a changing world. *Geographical Research*, 58(3). https://ur.booksc.me/book/83600795/a5c246

Váňa, J. (2020). Theorizing the social through literary fiction: For a new sociology of literature. *Cultural Sociology*, 14(2), 180–200.

Wadiwel, D. (2015). *The War Against Animals*. Leiden: Brill.

Watson, A. (2021). Writing sociological fiction. *Qualitative Research*. https://doi.org/10.1177/1468794120985677

Worrall, E. (2021). CNN: Australia is Shaping Up to be the Villain of COP26 Climate Talks. *Watts Up With That?* [BLOG], Sept 15, Chico, Newstex.

Index

Pages in *italics* refer to figures.

For Product Safety Concerns and Information please contact our EU
representative GPSR@taylorandfrancis.com
Taylor & Francis Verlag GmbH, Kaufingerstraße 24, 80331 München, Germany

www.ingramcontent.com/pod-product-compliance
Lightning Source LLC
Chambersburg PA
CBHW060302220326
41598CB00027B/4209

* 9 7 8 1 0 3 2 1 9 1 4 8 5 *